Higher BIOLOGY

PRACTICE QUESTION BOOK

Higher BIOLOGY PRACTICE QUESTION BOOK

John Di Mambro • Stuart White

001/30052018

10 9 8 7 6 5 4 3 2 1

ISBN 9780008263607

Published by

Leckie & Leckie Ltd

An imprint of HarperCollinsPublishers
Westerhill Road, Bishopbriggs, Glasgow, G64 2QT

T: 0844 576 8126 F: 0844 576 8131

leckieandleckie@harpercollins.co.uk www.leckieandleckie.co.uk

Special thanks to

Jouve (layout and illustration); Ink Tank (cover design);
Jess White (copy editing);

A CIP Catalogue record for this book is available from the British Library.

Printed and bound by CPI Group (UK) Ltd, Croydon, CR0 4YY

Check your own work. The answers are provided online at:

https://collins.co.uk/pages/scottish-curriculum-free-resources

How to use this book

Welcome to Leckie and Leckie's *Higher Biology Practice Question Book*. This book follows the structure of the Leckie and Leckie *Higher Biology Student Book*, so is ideal to use alongside it. Questions have been written to provide practice for topics and concepts which have been identified as challenging for many students.

Hints

Where appropriate, hints are provided to help give extra guidance and support.

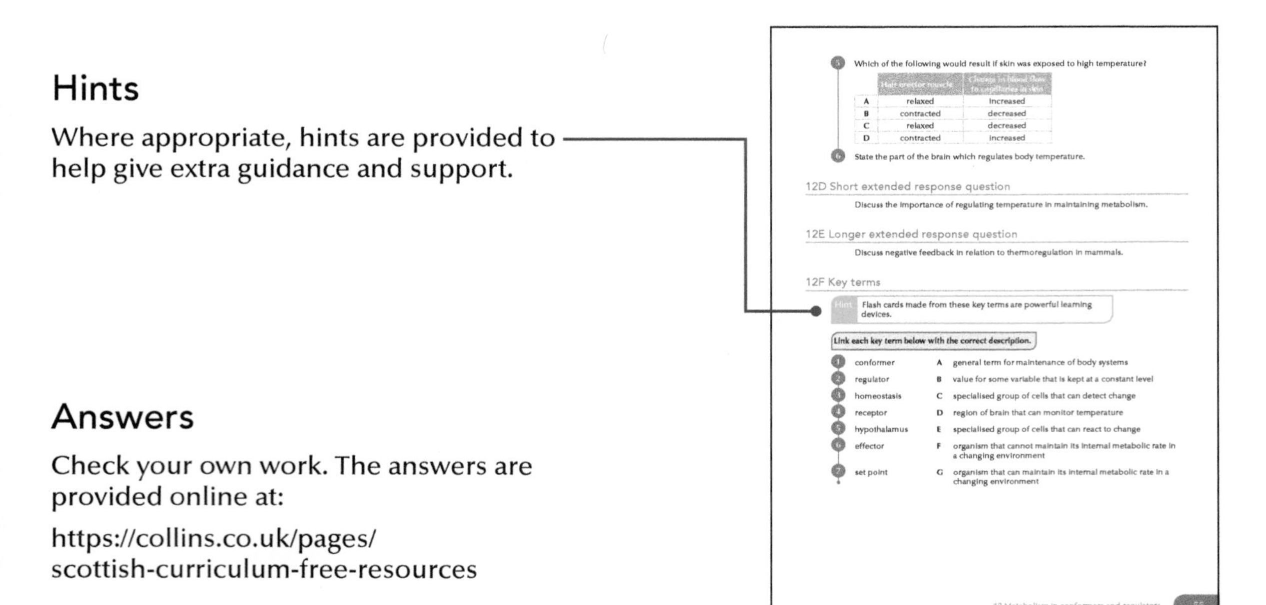

Which of the following would result if skin was exposed to high temperature?

	Hair erector muscle	Change in blood flow to capillaries in skin
A	relaxed	increased
B	contracted	decreased
C	relaxed	decreased
D	contracted	increased

State the part of the brain which regulates body temperature.

12D Short extended response question

Discuss the importance of regulating temperature in maintaining metabolism.

12E Longer extended response question

Discuss negative feedback in relation to thermoregulation in mammals.

12F Key terms

Hint Flash cards made from these key terms are powerful learning devices.

Link each key term below with the correct description.

1 conformer — A general term for maintenance of body systems
2 regulator — B value for some variable that is kept at a constant level
3 homeostasis — C specialised group of cells that can detect change
4 receptor — D region of brain that can monitor temperature
5 hypothalamus — E specialised group of cells that can react to change
6 effector — F organism that cannot maintain its internal metabolic rate in a changing environment
7 set point — G organism that can maintain its internal metabolic rate in a changing environment

12 Metabolism in conformers and regulators 55

Answers

Check your own work. The answers are provided online at:

https://collins.co.uk/pages/scottish-curriculum-free-resources

1 Structure of DNA

1A Structure of DNA

1. State what is meant by the genotype of a cell.

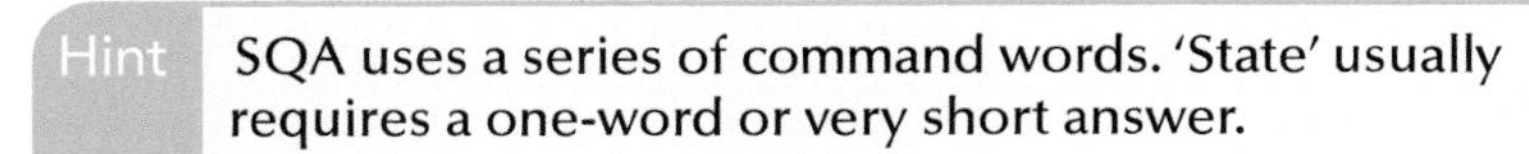

Hint SQA uses a series of command words. 'State' usually requires a one-word or very short answer.

2. Name the three components which make up a nucleotide.

3. Draw and label the nucleotide shown below.

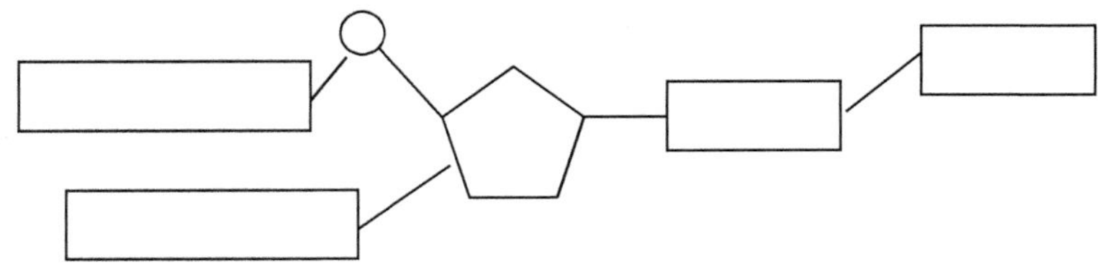

4. State how the two strands of DNA are held together.

Hint 'Describe' asks for more than a one-word answer. Instead, go further and make a statement.

5. Describe the shape of a DNA molecule.

6. Name the four nucleotide bases.

7. Explain what is meant by the following terms as applied to DNA structure:
   a complementary
   b antiparallel.

8. Draw two nucleotides. Join the 3′ carbon of the top nucleotide on to the 5′ carbon of the phosphate group of the next nucleotide. Now add a complementary strand.

Hint Use the format for drawing a nucleotide which is shown in Q3. Nucleotides are inverted on the opposite strand.

9. Decide if each of the following statements relating to DNA structure in the table below is **TRUE** or **FALSE**. If the statement is **FALSE**, write the correct term to replace the term underlined in the statement.

Hint When answering these questions, you must only replace the word underlined in the correction box when it is false. If you're unsure, go with 'true' to avoid the need to add a potentially wrong correction word.

Statement	True	False	Correction
DNA is composed of very long molecules made up of repeating chemical units called <u>bases</u>.			
Due to complementary base pairing, <u>adenine (A)</u> always pairs with thymine (T).			
A DNA strand has a <u>hydrogen</u> at its 3′ carbon end and a phosphate group at its 5′ carbon end.			

10 Copy and complete the following sentences by inserting one suitable term in each space provided.

A word may be used once, more than once or not at all.

chromosome – DNA – genetic – genotype – protein – sequence

All cells store their ________ information in the base sequence of ______.

The ___________is determined by the sequence of bases.

1B Organisation of DNA

1 Name the structure into which DNA is organised in the nucleus of a cell.

2 Give an example of a group of organisms composed of the following type of cells:

a eukaryotic

b prokaryotic.

3 State two organelles in eukaryotic cells which contain circular chromosomes.

4 Explain how a strand of DNA is organised to enable it to fit into a nucleus.

5 Copy and complete the table below to compare DNA organisation in prokaryotic and eukaryotic cells.

Hint 'Compare' means you demonstrate knowledge and understanding of the similarities and/or differences between topics being examined.

Cell characteristics	Prokaryote	Eukaryote
Contains nucleus (yes/no)		
Chromosomal DNA		
Plasmids in cytoplasm (yes/no)		

6 Decide if each of the following statements relating to DNA organisation in the table below is **TRUE** or **FALSE** and tick [√] the appropriate box. If the statement is **FALSE**, write the correct term in the correction box to replace the term <u>underlined</u> in the statement.

Statement	True	False	Correction
DNA is arranged in <u>linear</u> chromosomes and plasmids in prokaryotes.			
DNA is arranged in linear chromosomes in the nuclei of <u>eukaryotic</u> cells.			
DNA in linear chromosomes is tightly coiled and packaged with associated <u>glucose</u> in the nucleus of eukaryotes.			

7 Below are two types of cell with different DNA organisations.

cell P

cell Q

X

Y

nucleus (contains DNA)

Hint Familiarise yourself with diagrams of both prokaryotic and eukaryotic cells, but remember no two diagrams will be the same and the one in your exam is likely to be one you have not seen before.

a Give the letter of the prokaryotic cell.

b Identify structure X.

Hint Another command word. 'Identify' often means simply adding a label or a one-word answer.

c Describe how the DNA is organised in structures labelled Y.

d Name a group of eukaryotic organisms whose cells contain plasmids.

e Name an organelle, not shown above, in which DNA is organised in circular chromosomes.

8 The following are different structures into which DNA can be organised within cells.

1 Linear chromosome

2 Circular chromosome

3 Circular plasmid

Which of these terms describes how DNA is organised within bacterial cells?

A 1 only

B 2 only

C 1 and 2 only

D 2 and 3 only.

9 In which form does the genetic material of a eukaryote exist within the nucleus?

A Linear chromosomes

B Circular chromosomes

C Linear plasmids

D Circular plasmids.

1C Short extended response question

Write notes on the DNA organisation in prokaryotic and eukaryotic cells.

Hint This is typical of a question usually worth four or five marks in the Higher Examination, so aim to give at least four or five points, which must be relevant to the question.

1D Longer extended response question

Write notes on the structure of a DNA molecule.

> **Hint** This is typical of a question usually worth seven or more marks in the Higher Examination, so aim to give at least seven or more points, which must be relevant to the question.

1E Key terms

Link each key term below with the correct description.

> **Hint** If you are not sure of a term and its matching definition, then match up the ones you definitely know first.

	Key term		Description
1	genotype	**A**	parallel strands, but running in opposite directions
2	nucleotide	**B**	rule where A always pairs with T and C with G
3	double-stranded helix	**C**	adenine, thymine, cytosine, guanine
4	sugar–phosphate backbone	**D**	end of DNA strand with a deoxyribose sugar
5	nucleotide bases	**E**	holds base pairs (and the two strands) together
6	antiparallel	**F**	formed by joining nucleotides on the same strand
7	complementary base pairing	**G**	end of DNA strand with a phosphate
8	hydrogen bond	**H**	shape of a DNA molecule
9	3′ carbon end	**I**	made up of a deoxyribose sugar, phosphate and base
10	5′ carbon end	**J**	determined by the sequence of DNA bases in an organism
11	eukaryote	**K**	additional, circular DNA found in yeast and prokaryotes
12	prokaryote	**L**	allow DNA to be tightly coiled and packaged
13	linear chromosome	**M**	organism with a single cell which lacks a true membrane-bound nucleus
14	circular chromosome	**N**	DNA arrangement in nuclei of eukaryotic cells
15	plasmid	**O**	located in mitochondria and chloroplasts of eukaryotic cells
16	associated proteins	**P**	organism made of cells which contain a true membrane-bound nucleus

1F Data handling and experimental design

An experiment was set up to investigate the effect of temperature on sections of DNA with different numbers of A–T and C–G base pairs.

Hydrogen bonds join bases, with A–T having two and C–G three bonds. Each section of DNA was 100 base pairs long. The number of A–T base pairs was noted in each section before exposing the DNA to gradually increasing temperatures, until the strands separated. Seven strands were tested.

The results of the experiment are below.

Number of A–T base pairs in DNA section	Separation temperature (°C)
0	110
10	104
20	96
40	82
60	76
80	70
100	65

a **i** Name the dependent variable in this experiment.

Hint A dependent variable is usually what is measured, so the second column in a table.

ii Identify one variable, not already mentioned, that should be kept constant so that a valid conclusion can be drawn.

Hint Remember MTTVC – **M**ass, **T**ime, **T**emperature, **V**olume, **C**oncentration – this covers many of the variables used in experiments. And remember to say 'Mass of…', 'Volume of…', and on no account use the word 'amount'.

b Give a feature of the experiment that may make the results unreliable.

c Plot a line graph to show the results of the investigation.

d Draw a conclusion from these results.

e Suggest, with reference to number of A–T base pairs, why this conclusion was obtained.

Hint Another command word: 'Suggest' usually needs you to apply your knowledge to a new situation.

2 A section of double-stranded DNA was found to have 50 guanine bases and 20 adenine bases. Calculate the total number of deoxyribose sugars in this section.

> **Hint** 'Calculate' means you must determine a number from given facts, figures or information.

> **Hint** For questions like this, it's always helpful to write out the number of each base, so A = 20, G = 50, so T = __ and C = ___.
> Also remember DNA has two strands!

3 A double-stranded DNA fragment was found to have 180 nucleotides present. Which combination of bases would fit into this fragment?

A 40 thymine and 50 adenine

B 50 adenine and 50 cytosine

C 40 cytosine and 50 guanine

D 50 cytosine and 40 adenine

4 Part of a DNA molecule is 250 bases long. Of these bases, 60 are cytosine. The percentage of thymine bases present in this part of the DNA is:

A 12%

B 13%

C 24%

D 26%

2 Replication of DNA

2A Replication of DNA

1. State four substances required for DNA replication to occur.
2. Describe the function of the following in DNA replication:
   a primer
   b DNA polymerase
   c ligase.
3. Describe a difference in the replication of the two different strands of a DNA molecule.
4. Explain why cells need to carry out DNA replication.

Hint 'Explain' questions require more than straight recall of facts – you must provide a reason for the information given.

5. State the purpose of polymerase chain reaction (PCR).
6. Name one practical application of PCR.
7. Copy and complete the table, describing what is happening at each stage of PCR.

Temperature (°C)	Description of events
90	
55	
72	

8. Decide if each of the following statements relating to DNA replication and PCR in the table below is **TRUE** or **FALSE.** If the statement is **FALSE**, write the correct term to replace the term underlined in the statement.

Statement	True	False	Correction
Ligase adds complementary nucleotides to the deoxyribose (3′) end of a DNA strand.			
DNA polymerase can only add DNA nucleotides in one direction resulting in one strand being replicated continuously and the other strand replicated in fragments.			
Cooling allows primers to bind to target sequences.			

9 The diagram shows a strand of DNA during replication and associated structures.

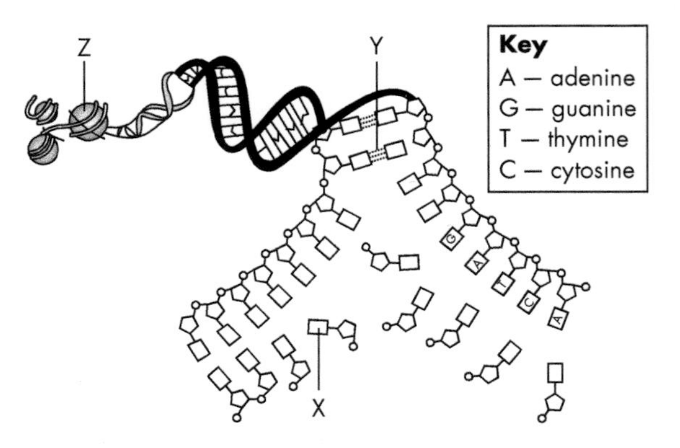

a Name base X.

b Name bond Y.

c State the role of molecule Z.

d Explain the function of a primer.

e State the name of the enzyme that joins fragments of DNA together.

10 Which of the following are required in PCR?

A DNA polymerase, template strand and primers

B RNA polymerase, template strand and primers

C DNA polymerase, template strand and ligase

D RNA polymerase, ligase and primers

11 Each cycle of a particular PCR takes 10 minutes.

If there are 5000 DNA fragments at the start of the reaction, how long will it take for the number of fragments produced by the reaction to be greater than one million?

A 70 minutes

B 80 minutes

C 90 minutes

D 200 minutes

12 Which of the following molecules are required in the continuous replication of a strand of a DNA molecule?

A DNA polymerase and ligase only

B DNA polymerase and primers only

C Ligase and primers only

D DNA polymerase, ligase and primers

13 The following are steps in the process of PCR.

1 Cooling allows primers to bind to target sequences.

2 DNA is heated to separate the strands.

3 Repeated cycles of heating and cooling amplify this region of DNA.

4 Heat-tolerant DNA polymerase replicates the region of DNA.

Choose the letter that puts the above steps into the correct order.

A $1 \rightarrow 2 \rightarrow 3 \rightarrow 4$

B $2 \rightarrow 1 \rightarrow 3 \rightarrow 4$

C $2 \rightarrow 1 \rightarrow 4 \rightarrow 3$

D $4 \rightarrow 2 \rightarrow 3 \rightarrow 1$

14 DNA polymerase:

A causes the formation of hydrogen bonds between bases

B cannot add nucleotides to an already existing DNA chain

C requires a primer to be present

D can only add nucleotides to the free 5′ carbon end of a DNA strand.

15 The graph below shows the temperature changes involved in one cycle of PCR.

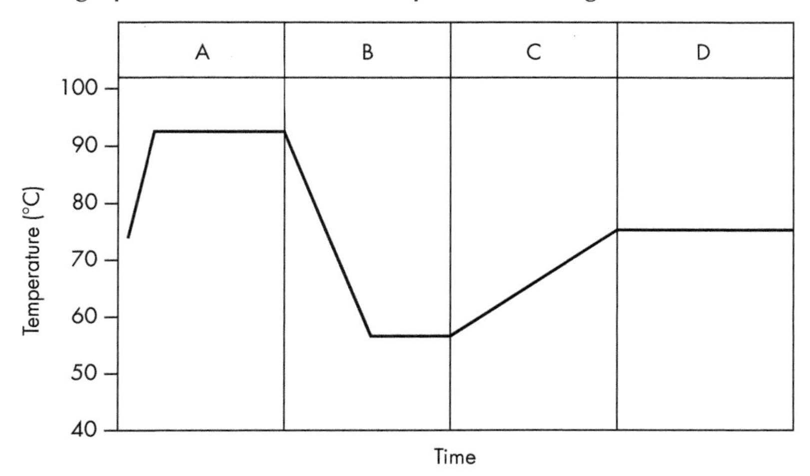

Which letter indicates when primers would bind to target sequences of DNA?

16 Copy and complete the following sentences by inserting one suitable term in each space provided.

A word may be used once, more than once or not at all.

bases – continuously – locations – unwound – fragments – sequence – replicated – primer – nucleotides

Prior to cell division, DNA is ________ by a DNA polymerase. DNA polymerase needs a ________ to start replication.

DNA is ________ and unzipped to form two template strands. This process occurs at several ___________ on a DNA molecule. DNA polymerase can only add DNA ___________ in one direction resulting in one strand being replicated ____________ and the other strand replicated in ___________.

2B Short extended response question

Give an account of the roles of enzymes in DNA replication and PCR.

2C Longer extended response question

1 Write notes on:

i DNA replication

ii amplification of DNA.

2D Key terms

Link each key term below with the correct description.

	Key term		Description
1	DNA replication	**A**	process of copying a strand of DNA nucleotide bases
2	primer	**B**	will not denature in high temperatures
3	DNA polymerase	**C**	high-energy molecule synthesised in respiration
4	ATP	**D**	sequence of nucleotide bases which are complementary to a specific primer
5	ligase	**E**	small strand of DNA required for replication to start
6	leading strand	**F**	strand being replicated in fragments
7	lagging strand	**G**	strand being replicated continuously
8	PCR	**H**	joins fragments of DNA together
9	target sequence	**I**	adds complementary nucleotides to the deoxyribose (3′) end of a DNA strand
10	heat tolerant	**J**	technique for the amplification of DNA *in vitro*

2E Data handling and experimental design

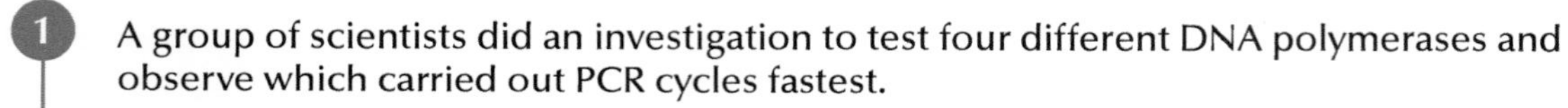

1. A group of scientists did an investigation to test four different DNA polymerases and observe which carried out PCR cycles fastest.

   They carried out this reaction by adding 5 µl polymerase buffer, 5 µl of each primer, 1.0 µl DNA polymerase and the DNA template.

   They measured the time for 30 cycles, for each DNA polymerase, and calculated the average time for one cycle (mins). The results are shown in the table below.

DNA polymerase	Average time for one cycle (mins)
1	5
2	10
3	18
4	26

a **i** Name the dependent variable in this experiment.

ii Identify one variable, not already mentioned, that should be kept constant so that a valid conclusion can be drawn.

b Give a feature of the experiment which made the results more reliable.

Hint Reliability is increased by repeating the investigation for each independent variable.

c Explain the importance of using heat-tolerant DNA polymerases in PCR.

d Draw a conclusion from these results.

e Suggest why the scientists might want to investigate different DNA polymerases.

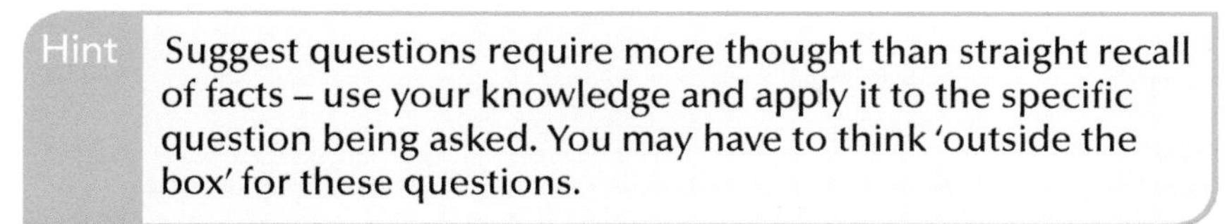

Hint Suggest questions require more thought than straight recall of facts – use your knowledge and apply it to the specific question being asked. You may have to think 'outside the box' for these questions.

The tube below shows the mixture of chemicals used in the PCR process.

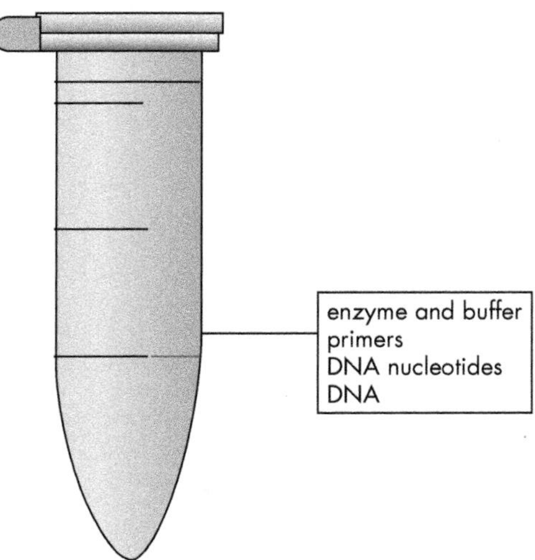

f Describe the contents of a control tube designed to show the need for enzyme to be present.

3 Gene expression

3A Structure and function of RNA

1 Complete the table below to show the differences between DNA and RNA.

	DNA	RNA
Where found in the cell		
Number of strands		
Type of sugar in nucleotide		
Adenine paired to		

Hint: Comparing DNA and RNA is a common question. In an open question, be sure to compare them, i.e. DNA is … but RNA is ….

2 Copy and complete the table below to state the location and function of different RNA forms.

	Location	Function
mRNA		
tRNA		
rRNA		

3 Information about DNA, mRNA and tRNA is given below.

Which row in the table is correct?

	DNA	mRNA	tRNA
A	Double-stranded Deoxyribose sugar present	Single-stranded Contains codons	Double-stranded Contains codons
B	Single-stranded Ribose sugar present	Single-stranded Contains anti-codons	Single-stranded Contains anti-codons
C	Single-stranded Deoxyribose sugar present	Single-stranded Contains anti-codons	Double-stranded Contains codons
D	Double-stranded Deoxyribose sugar present	Single-stranded Contains codons	Single-stranded Contains anti-codons

3B Transcription

1. Describe what is meant by 'gene expression'.
2. Name the enzyme involved in transcription.
3. State the location of transcription.
4. Name the coding and non-coding regions of a primary transcript of mRNA.
5. Describe RNA splicing of a primary mRNA transcript.

3C Translation

1. Name the organelle in which translation takes place.
2. Name the triplet of bases on tRNA which codes for a specific amino acid.
3. Name the bonds which join amino acids into a chain.
4. Use information in the diagram to answer the following.
   a Name the molecule shown in the diagram.
   b Name regions X and Z.
   c Name bond Y.
   d Describe the function of this molecule.

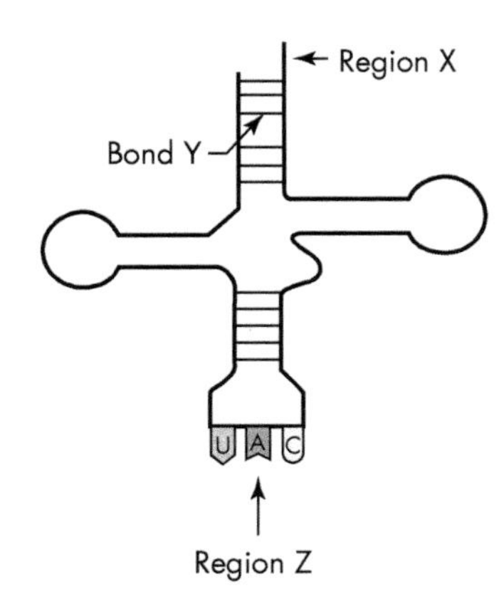

3D One gene – many proteins

1. Describe how protein structure and shape can vary.
2. The diagram shows the steps during transcription and translation of a gene.

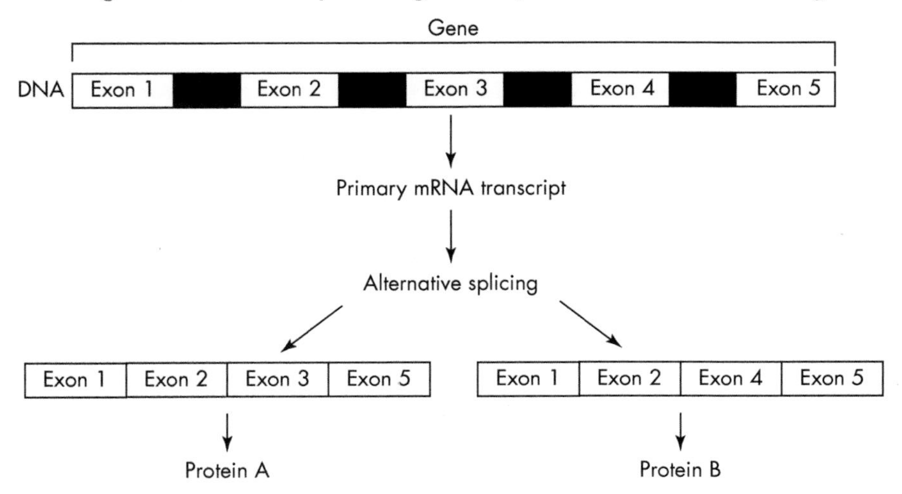

   a State all the different combinations of exons that could be spliced together to produce different proteins from those shown above.
   b Name the regions shown in black.

3 Decide if each of the following statements relating to gene expression in the table below is **TRUE** or **FALSE**. If the statement is **FALSE**, write the correct term to replace the term underlined in the statement.

Statement	True	False	Correction
Amino acids are linked by hydrogen bonds to form polypeptides.			
RNA has a single strand, replaces thymine with uracil and deoxyribose with ribose.			
Each rRNA carries a specific amino acid.			
The introns of a mature transcript of mRNA are non-coding and are removed in RNA splicing.			
tRNA folds due to base pairing to form a triplet codon site and an attachment site for a specific amino acid.			
Different mRNA molecules are produced from the same primary transcript depending on which exons are included in the mature RNA transcript.			

4 During transcription, which regions of a primary transcript are non-coding and removed in RNA splicing?

A Codons

B Anti-codons

C Introns

D Exons

5 Gene expression is controlled by regulating:

A translation only

B transcription only

C both translation and transcription

D post-translational modification.

6 Copy and complete the following sentences by inserting one suitable term in each space provided.

A word may be used once, more than once or not at all.

peptide – structures – hydrogen – functions – fold – amino

Proteins have a large variety of __________ and shapes resulting in a wide range of __________.

Amino acids are linked by __________ bonds to form polypeptides. Polypeptide chains ______ to form the three-dimensional shape of a protein, held together by __________ bonds and other interactions between individual __________ acids.

3E Short extended response question

Write notes on the differences between RNA and DNA and the different forms of RNA.

3F Longer extended response question

1 Give an account of protein synthesis under the following headings:

i transcription

ii translation

3G Key terms

Link each key term below with the correct description.

	Key term		Description
1	RNA	**A**	amino acids linked by peptide bonds
2	uracil	**B**	non-coding regions of mRNA removed in RNA splicing
3	ribose	**C**	mRNA with introns removed and exons joined together
4	mRNA	**D**	moves along DNA unwinding and unzipping the double helix and synthesises a primary transcript of RNA from RNA nucleotides by complementary base pairing
5	tRNA	**E**	carries a specific amino acid
6	rRNA	**F**	triplet of nucleotide bases on mRNA
7	gene expression	**G**	coding regions of mRNA which make up the mature transcript
8	transcription	**H**	using mRNA to produce a polypeptide by tRNA at the ribosome
9	translation	**I**	triplet of nucleotide bases on tRNA
10	intron	**J**	mRNA produced by complementary base pairing of RNA nucleotides
11	exon	**K**	nucleotide base in RNA instead of thymine
12	primary transcript	**L**	combines with protein to form the ribosome
13	mature transcript	**M**	sugar found in RNA instead of deoxyribose
14	RNA splicing	**N**	linked by peptide bonds to form polypeptides
15	RNA polymerase	**O**	joins amino acids to form a polypeptide
16	codon	**P**	using DNA to produce primary/mature RNA transcripts
17	anti-codon	**Q**	location of translation
18	amino acid	**R**	single-stranded molecule similar to DNA
19	peptide bond	**S**	the removal of introns from a primary transcript and joining together of exons to produce a mature transcript
20	polypeptide chain	**T**	process by which a gene is used to direct protein synthesis
21	ribosome	**U**	carries a copy of the DNA code from the nucleus to the ribosome

Electrophoresis is a technique that is used to separate large molecules such as proteins or DNA.

Electrophoresis can be used to compare proteins from different sources. Protein samples are loaded into the sample well and when a current is applied, the proteins will travel through the gel. The smaller the size of the protein (kD), the further it travels through the gel.

The set-up is shown below.

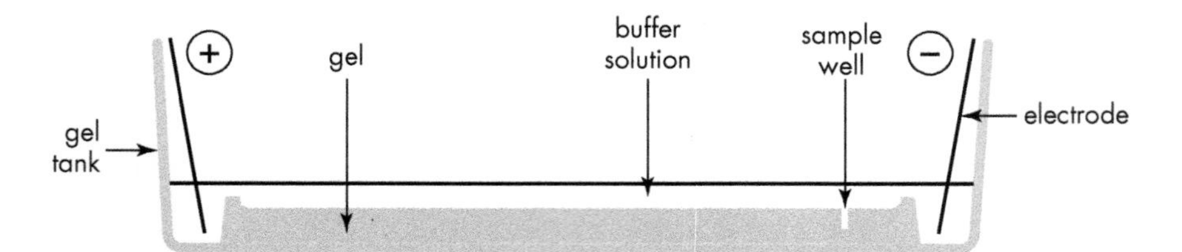

Students carried out an experiment to separate proteins by size using this method.

The results from five different protein samples they tested are shown in the table below.

Protein	Size (kD)
Myosin	205
Actin	42
Rubisco	55
Casein	23
Egg albumin	45

a Name the dependent variable in this experiment.

b Identify one variable, not already mentioned, that should be kept constant so that a valid conclusion can be drawn.

c Give a feature of the experiment that may make the results unreliable.

d Plot a line graph to show the results of the investigation.

e Draw a conclusion from these results.

> **Hint** Conclusions must involve the dependent and independent variables for the experiment. Refer back to the aim. Sometimes the dependent variable is measured by an indicator, but it's the dependent variable you would mention in your answer.

f Suggest why a buffer solution is added to the gel.

4 Cellular differentiation

4A Cellular differentiation

1. State what is meant by the term cellular differentiation.
2. State where differentiation occurs in:
   a a plant
   b an animal.
3. State one part of the adult human body where stem cells are found.
4. Name two uses of stem cells in research.

> Hint: This is commonly not answered well in tests and exams. Be sure to learn this.

5. Name two therapeutic uses of stem cells.
6. Decide if each of the following statements relating to cellular differentiation in the table below is **TRUE** or **FALSE**. If the statement is **FALSE**, write the correct term to replace the term underlined in the statement.

Statement	True	False	Correction
Cellular differentiation is the process by which a cell develops more specialised functions by expressing the genes characteristic for that type of cell.			
Stem cells can be used as model cells to study how diseases develop or for drug testing.			
Embryonic stem cells replace differentiated cells that need to be replaced and give rise to a more limited range of cell types.			
Stem cells are regions of unspecialised cells in plants that are capable of cell division.			

7 The diagram below shows differentiation of a stem cell.

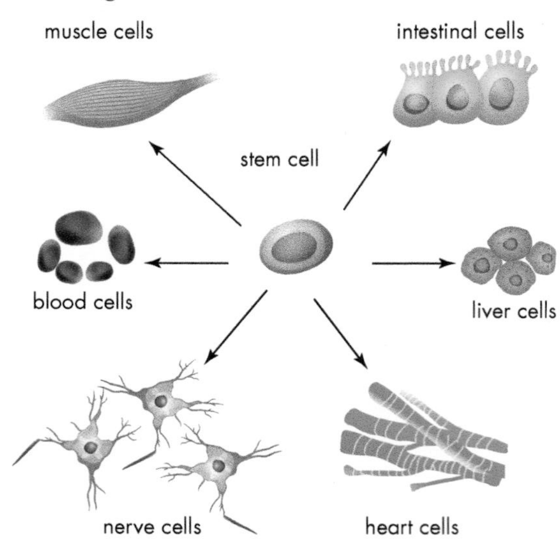

a Explain how the diagram above indicates that the stem cell shown is an embryonic and not a tissue (adult) stem cell.

b After a stem cell differentiates, only certain genes are expressed.

Explain how this results in different cell types.

8 Copy and complete the following sentences by inserting one suitable term in each space provided.

A word may be used once, more than once or not at all.

divide – unspecialised – stem – differentiate – meristems – specialised – multiply

________ are regions of ________ cells in plants that are capable of cell division.

________ cells are unspecialised somatic cells in animals that can ________ to make copies of themselves (self-renew) and/or ________ into specialised cells.

4B Short extended response question

1 Give an account of:

i ethical issues associated with the use of stem cells

ii therapeutic and research uses of stem cells.

4C Longer extended response question

Write notes on the differentiation of stem cells.

4D Key terms

Link each key term below with the correct description.

	Key term		Description
1	differentiation	A	provides information on how cell processes such as cell growth, differentiation and gene regulation work
2	meristem	B	repair of damaged or diseased organs, e.g. corneal transplants, and skin grafts for burns
3	stem cell	C	unspecialised cells in the very early embryo
4	pluripotent	D	unspecialised cells found in the tissues of an organ
5	multipotent	E	used to study how diseases develop or for drug testing
6	therapeutic uses of stem cells	F	regions of unspecialised cells in plants that can divide (self-renew) and/or differentiate.
7	embryonic stem cell	G	can differentiate into all the cell types that make up the organism
8	tissue stem cell	H	concerns arising from a set of principles regarding moral values and appropriate conduct, e.g. destruction of embryos used in research
9	stem cell research	I	unspecialised somatic cells in animals that can divide to make copies of themselves (self-renew) and/or differentiate into specialised cells
10	model cells	J	give rise to a more limited range of cell types
11	ethical issues	K	process by which a cell develops more specialised functions by expressing the genes characteristic for that type of cell

Meristems can be cultured in a growth medium to produce new plants.

An experiment was carried out to investigate the effects of four different growth media (A, B, C and D) on the production of shoots by meristems of plants.

Three meristems were removed and cultured in each medium for a period of ten weeks. The average number of shoots produced per meristem was recorded at specific times during the investigation.

The results are shown in the table below.

Week	Average number of shoots produced per meristem (for each growth medium)			
	A	B	C	D
1	0	2	3	1
2	1	3	5	2
3	1	5	7	3
4	1	6	9	5
5	3	7	11	7
6	4	8	11	10
7	5	10	11	14
8	5	12	11	16
9	5	13	11	17
10	5	13	11	18

a **i** Name the independent variable in this experiment.

Hint: The independent variable is the one changed by the investigator.

ii Identify one variable, not already mentioned, that should be kept constant so that a valid conclusion can be drawn.

b Give a feature of the experiment which may make the results unreliable.

c Plot a line graph to show the results of growth medium D.

d Predict the average number of shoots produced using growth medium D after 12 weeks, assuming the same rate of shoot production.

Hint: Another command word. 'Predict' asks you to work out what will happen based on the information available.

e A food production scientist can use the shoots grown from the meristems to plant and grow crops.

Suggest which growth medium would be most suitable:

i in a country with a very short growing season

ii to maximise overall yield.

5 Structure of the genome

5A Structure of the genome

1. State what is meant by the term genome.
2. Name the two types of sequences within the genome.
3. Name the molecules that are coded for by genes.
4. Name two types of RNA that are transcribed but not translated.

> **Hint** This is also often poorly answered in tests and exams. Be sure to learn these.

5. The genome of an organism is its hereditary information encoded in the:

   A ribosome

   B cell membrane

   C cytoplasm

   D DNA.

6. Which of the following is incorrect?

   The human genome contains base sequences:

   A that regulate translation

   B that regulate transcription

   C that are transcribed to RNA but never translated

   D from which primary transcripts are produced.

5B Short extended response question

Describe the structure of the genome.

5C Key terms

Link each key term below with the correct description.

	Key term		Description
1	genome	**A**	DNA sequences that code for protein
2	genes	**B**	sequences including genes
3	coding sequences	**C**	sequences that regulate transcription and those that are transcribed to RNA but are never translated, or have no known function
4	non-coding sequences	**D**	entire hereditary information encoded in DNA of an organism

6 Mutations

6A Single gene mutations

1. Define the term mutation.
2. State what is meant by a single gene mutation.
3. Describe how the DNA nucleotide sequence is altered in the following mutation types:
   - **a** substitution
   - **b** insertion
   - **c** deletion.
4. The table below shows a DNA base sequence with no mutations and three mutated sequences.
   - **a** Name the types of mutation A, B and C.
   - **b** State whether the impact of each mutation, A, B and C is major or minor.

Mutation	DNA base sequence
None	ATG CGT CGA
A	ATG CTC GAA
B	ATG CGG TCG
C	ATG CCT CGA

5. Describe what is meant by the term frame-shift mutation.
6. Explain the consequences of nucleotide insertions or deletions in terms of the production of proteins.
7. Copy and complete the table below.

Substitution mutation type	Change	End effect on protein produced

8. The following is a sequence of DNA bases.

   T–G–C–A–A–G–C–G–T

   The sequence of bases is altered by a mutation and is changed to the sequence below.

   T–G–C–A–A–C–C–G–T

   Which type of mutation has occurred?

   A Deletion

   B Insertion

   C Substitution

   D Duplication

9. A missense point mutation:

   A is a substitution and the protein formed functions in a different way

   B brings the translation process to a complete halt

   C causes all the subsequent codons to be altered

   D does not affect the amino acid coded for by the altered codon.

6B Chromosome structure mutations and the importance of mutations and gene duplication in evolution

1. Describe what is meant by a chromosome structure mutation.
2. Describe how the chromosome structure is altered in the following mutation types:

   a duplication

   b inversion

   c deletion

   d translocation.

   > **Hint** Be sure not to mix up the chromosome structure and single gene mutations. Deletion is common to both but has different definitions.

3. Explain why single gene mutations are important in terms of evolution.
4. Explain the effect of mutation in duplicated genes.

6C Longer extended response question

Give an account of single gene and chromosome structure mutations.

6D Key terms

Link each key term below with the correct description.

1. mutation
2. single gene mutation
3. substitution
4. insertion
5. deletion
6. frame-shift mutation
7. missense
8. nonsense
9. splice site
10. chromosome structure mutation
11. duplication
12. inversion
13. translocation

A a section of a chromosome is added from its homologous partner

B substitution that results in some introns being retained and/or some exons not being included in the mature transcript

C causes all of the codons and all of the amino acids after the mutation to be changed. This has a major effect on the structure of the protein produced

D substitution that results in one amino acid being changed for another. This may result in a non-functional protein or have little effect on the protein

E a section of chromosome is reversed

F base replaced by another, with no other bases changing

G base deleted from the sequence/gene removed from chromosome

H a section of a chroosome is added from its homologous partner

I changes in the genome that can result in no protein or an altered protein being expressed

J additional base inserted into the sequence

K substitution that results in a premature stop codon being produced, which results in a shorter protein

L alteration of a DNA nucleotide sequence

M alterations to the structure of one or more chromosomes

6E Data handling and experimental design

The table below shows the results of an experiment where two different yeast cultures, A and B, are exposed to UV light for 24 hours. The two cultures, A and B, are grown and form one hundred colonies (groups of cells) before the start of the experiment. Every 4 hours, the number of colonies with viable yeast was counted in each culture.

Exposure time (hours)	Yeast culture colonies	
	A	B
0	100	100
4	82	96
8	64	95
12	47	94
16	37	81
20	24	79
24	12	79

a State the yeast culture, A or B, that is most sensitive to UV light.

b Name two variables, not mentioned above, that would have to be kept constant during this experiment.

c State one way in which the reliability of these results could be improved.

d Plot a line graph to show the results of the investigation.

e Draw a conclusion from these results.

7 Evolution

7A Evolution, gene transfer and selection

1. State what is meant by the term evolution.
2. Describe what is meant by the following types of gene transfer:
   a vertical **b** horizontal.
3. Name two groups of organisms that can transfer genes horizontally.
4. Explain what is meant by natural selection.
5. Describe the different outcomes as a result of the types of selection below.
   a Stabilising **b** Directional **c** Disruptive

Hint: Remember these are the outcomes of natural selection, not mechanisms by which selection occurs.

6. The graphs below show possible changes in the height of organisms in a population in response to a selection pressure.

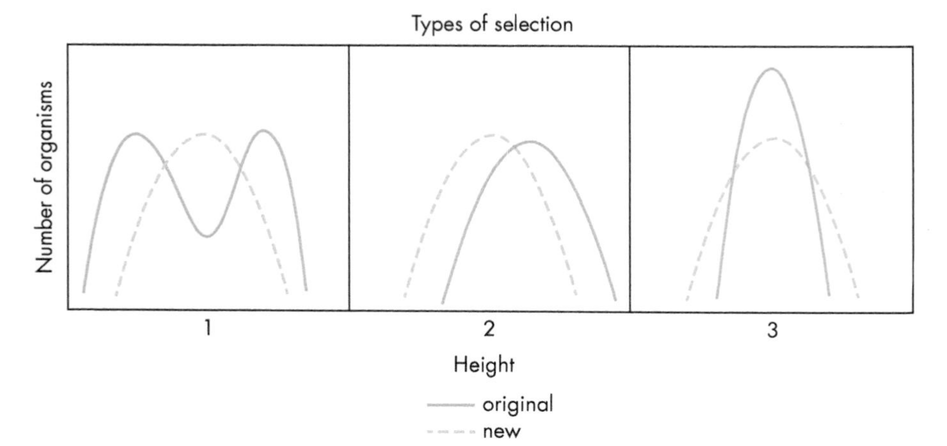

Which row in the table below matches each graph with the type of selection having taken place?

	Graph		
	1	2	3
A	disruptive	directional	stabilising
B	directional	disruptive	stabilising
C	stabilising	disruptive	directional
D	directional	stabilising	disruptive

7B Speciation

1. State what is meant by the term species.
2. Describe what is meant by the term speciation.
3. State the importance of isolation barriers in speciation.
4. Briefly describe the three stages of speciation.
5. Describe what is meant by the term allopatric speciation.
6. State three examples of a geographical barrier.
7. Briefly describe what is meant by the term sympatric speciation.
8. Decide if each of the following statements relating to evolution in the table below is **TRUE** or **FALSE**. If the statement is **FALSE**, write the correct term to replace the term <u>underlined</u> in the statement.

Statement	True	False	Correction
<u>Evolution</u> is defined as the changes in organisms over generations as a result of genomic variations.			
Prokaryotes and viruses can transfer sequences <u>vertically</u> into the genomes of eukaryotes.			
<u>Sexual</u> selection is the non-random increase in frequency of DNA sequences that increase survival and the non-random reduction in harmful sequences.			
Genetic drift has a greater impact in <u>small</u> populations.			
Speciation is the generation of new biological species by evolution as a result of <u>substitution</u>, mutation and selection.			

9. Which of the following terms describes the generation of new biological species by evolution?

   A Selection
   B Speciation
   C Gene transfer
   D Genetic drift

7C Short extended response question

1. Give an account of speciation under the following headings:

   i allopatric speciation
   ii sympatric speciation.

7D Longer extended response question

1. Write notes on evolution under the following headings:

   i natural selection
   ii genetic drift.

7E Key terms

Link each key term below with the correct description.

	Key term		Description
1	evolution	A	group of organisms capable of interbreeding and producing fertile offspring, and which does not normally breed with other groups
2	vertical gene transfer	B	form of natural selection that favours an extreme characteristic away from the median characteristics
3	horizontal gene transfer	C	evolutionary process by which new species are formed due to an ecological or behavioural barrier
4	natural selection	D	non-random increase in frequency of DNA sequences that increase survival and the non-random reduction in harmful sequences
5	stabilising selection	E	generation of new biological species by evolution as a result of isolation, mutation and selection
6	directional selection	F	method used by prokaryotes to transfer genes from one cell to another
7	disruptive selection	G	form of natural selection that favours the median characteristics in variation
8	species	H	evolutionary process where new species are formed due to a geographical barrier
9	speciation	I	when genes are transferred from parent to offspring as a result of sexual or asexual reproduction
10	allopatric speciation	J	form of natural selection that favours two extreme characteristics at the expense of the median characteristics
11	sympatric speciation	K	changes in organisms over generations as a result of genomic variations

7F Data handling and experimental design

Students carried out an investigation to see how frequently two different species of bird, capable of cross breeding, were present in a hybrid zone.

They measured the number of each species that were present within the hybrid zone between January and June. They observed the zone for 1 hour each day during these months. They recorded the data in the table below.

Month observed	Number of days (per month) species spotted in hybrid zone		
	Species 1	Species 2	Both present
January	10	0	0
February	16	0	0
March	25	20	18
April	30	30	30
May	30	14	14
June	15	4	3

a **i** Name the dependent variable in this experiment.

ii Identify one variable, not already mentioned, that should be kept constant so that a valid conclusion can be drawn.

b Give a feature of the experiment which may make the results unreliable.

c Suggest how the students could improve the reliability of the investigation.

d Suggest why none of species 2 were spotted in January or February.

e Use the data in the table above to answer the following.

i Calculate the percentage increase in the number of days only species 1 was spotted between January and April.

Hint: Percentage change is calculated by dividing the change by the original value and multiplying by 100.

ii State the simplest whole number ratio between the number of days only species 1 and only species 2 were spotted in March.

iii Calculate the average increase in the number of days species 1 was spotted between January and May.

Hint: Average increase is calculated by taking the total increase and dividing that number by the time period, in this instance, months.

8 Genomic sequencing

8A Genomic sequencing

1. State the definition of the term genomic sequencing
2. Explain why genomic sequencing is useful.
3. **a** State the term used to describe the analysis of sequence data using computers and statistics.
   b State the advantage of using this method of analysis.
4. State what is meant by the term phylogenetics.
5. State how phylogenetic trees and molecular clocks determine the main sequence of events in evolution.
6. The study of evolutionary relatedness among different groups of organisms can be used to produce diagrams as such shown opposite and below.
   a **i** Name this area of study.
   ii From the diagram opposite, give the letters of the two organisms that are the most closely related.

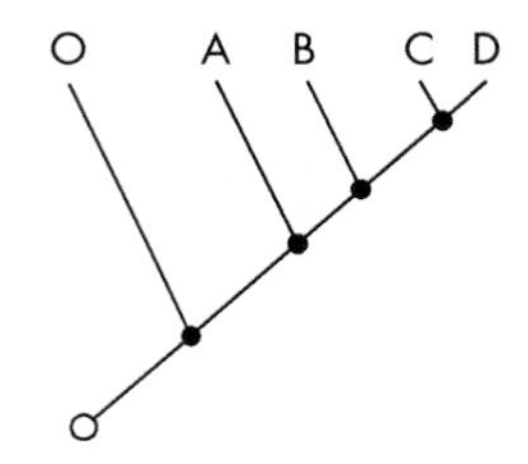

   b The diagram below shows a molecular clock for some mammals.

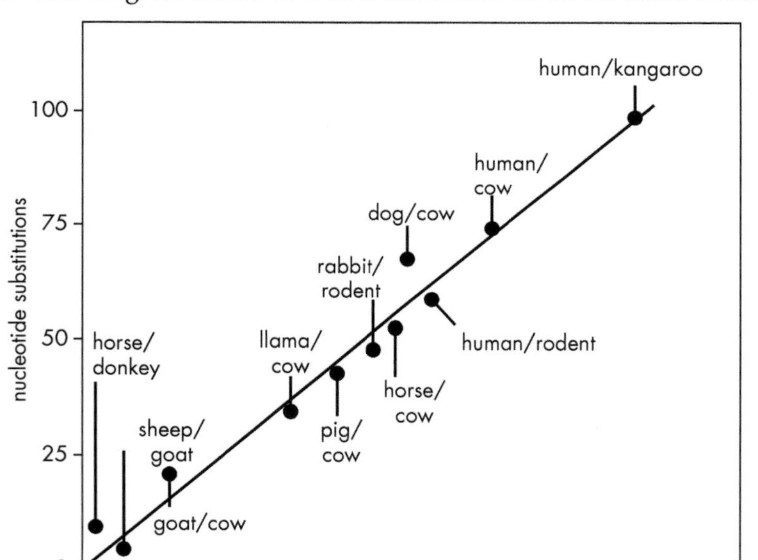

   i Describe what this molecular clock shows.
   ii According to the clock above, name the two pairs of organisms that are genetically closest to a common ancestor.

7 State two sources of evidence used to determine the main sequence of events in evolution.

8 Name the three domains of life.

9 State what scientists have concluded when comparing genomes in different organisms.

10 State what is meant by the term personal genomics.

11 State what is meant by the term pharmacogenetics.

12 Describe two of the main difficulties with personalised medicine.

13 The diagram below shows a phylogenetic tree.

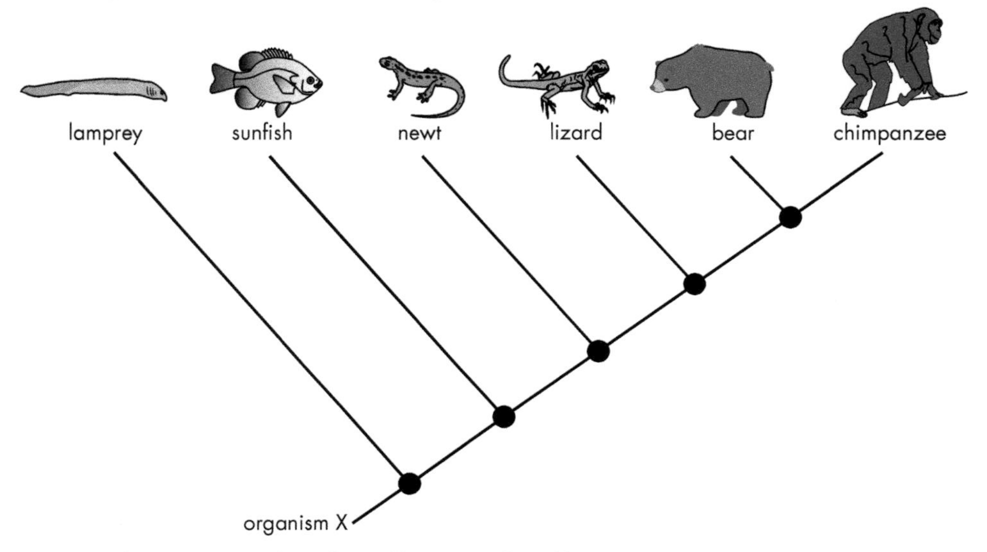

a State the term used to describe organism X.

b Name the two species most closely related in this tree.

c Bears are more closely related to lizards than they are to newts.
Use evidence from the phylogenetic tree to justify this statement.

14 Decide if each of the following statements relating to genomic sequencing in the table below is **TRUE** or **FALSE**. If the statement is **FALSE**, write the correct term to replace the term underlined in the statement.

Statement	True	False	Correction
Phylogenetics is the study of evolutionary relatedness among groups of organisms.			
The sequence of events in evolution can be determined using species data and fossil evidence.			
Comparison of genomes reveals that many genes are rarely conserved across different organisms.			
Bioinformatics is used to compare sequence data, using computer and statistical analyses.			

15 The diagram below represents a phylogenetic tree showing evolutionary relatedness of different species.

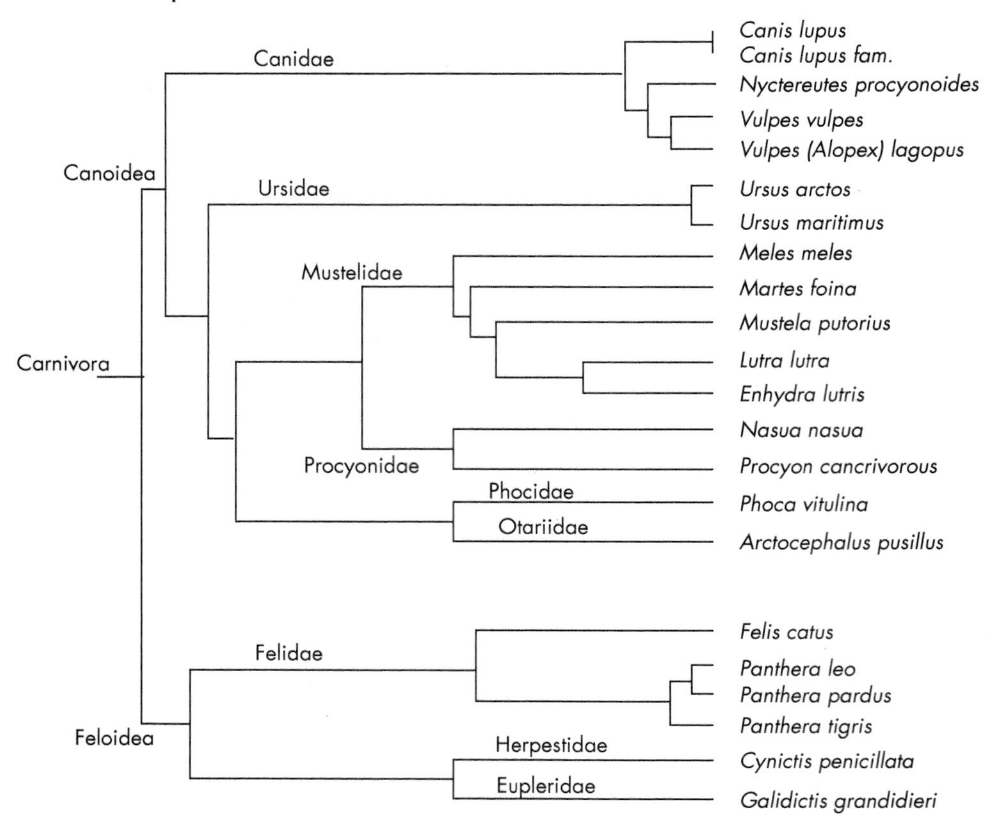

With how many species does the *Lutra lutra* share a common ancestor in this phylogenetic tree?

A 4

B 15

C 21

D 22

16 The study of how groups of organisms are related through evolution is called:

A pharmacogenetics

B phylogenetics

C gene sequencing

D bioinformatics.

17 Which of the following correctly describes the main sequence of events in the evolution of life, with the earliest first?

A Photosynthetic organisms, eukaryotes, multicellular organisms, vertebrates

B Vertebrates, photosynthetic organisms, eukaryotes, multicellular organisms

C Photosynthetic organisms, multicellular organisms, eukaryotes, vertebrates

D Vertebrates, eukaryotes, multicellular organisms, photosynthetic organisms

8B Short extended response question

Give an account of personalised genomics and medicine.

8C Longer extended response question

Give an account of phylogenetics and molecular clocks.

8D Key terms

Link each key term below with the correct description.

	Key term		Description
1	genomic sequencing	A	bacteria, archaea and eukaryotes
2	bioinformatics	B	study of evolutionary relatedness of species
3	sequence data	C	study of inherited genetic differences in drug metabolic pathways which can affect individual responses to drugs, both in terms of therapeutic and adverse effects
4	fossil evidence	D	DNA sequences found across many species
5	phylogenetics	E	determining the sequence of nucleotide bases for individual genes and entire genomes
6	molecular clocks	F	analysis of sequence data using computers and statistics
7	universal ancestor	G	non-human species that is extensively studied to provide insight into the workings of other organisms
8	common ancestor	H	concerned with the sequencing, analysis and interpretation of the genome of an individual
9	three domains of life	I	used to determine the sequence of events in evolution
10	model organisms	J	treatment which is based upon an individual's own genome
11	conserved genes	K	most recent ancestral form or species from which two different species evolved
12	personal genomics	L	most recent population of organisms from which all organisms now living on Earth have a common descent
13	pharmacogenetics	M	graph that shows differences in sequence data for nucleic acids or proteins over time
14	personalised medicine	N	used to study the evolutionary relatedness among groups of organisms

The table below shows the estimated comparative genome sizes of humans compared to other model organisms.

Organism	Estimated size (base pairs)	Chromosome number	Estimated gene number
Human (*Homo sapiens*)	3 billion	46	25 000
Mouse (*Mus musculus*)	2.9 billion	40	25 000
Fruit fly (*Drosophila melanogaster*)	165 million	8	13 000
Thale cress (*Arabidopsis thaliana*)	157 million	10	25 000
Roundworm (*Caenorhabditis elegans*)	97 million	12	19 000
Yeast (*Saccharomyces cerevisiae*)	12 million	32	6000
Bacteria (*Escherichia coli*)	4.6 million	1	3200

a State the relationship between estimated genome size and estimated gene number.

b Calculate the percentage increase in the chromosome number between thale cress and a mouse.

c State the simplest whole number ratio for estimated genome size between humans and yeast.

d Calculate the average number of genes on each chromosome in a fruit fly.

e Suggest a reason for the gene number of the fruit fly being lower than for thale cress, despite having a greater estimated genome size.

9 Metabolism and survival

9A Introduction to metabolic pathways

1. Describe what is meant by the term metabolism.

> **Hint** SQA uses a series of command words. 'Describe' asks for more than a one-word answer. Instead, go further and make a statement.

2. Copy and complete the following sentences describing metabolic pathways by inserting one suitable term in each space provided

   A word may be used once, more than once or not at all.

control – enzymes – metabolic – integrating – inhibit – proteins

   In living systems, many ___________ pathways operate at the same time. All such pathways are catalysed by ___________ which help regulate or ___________ metabolism. In order for the hundreds of different pathways to function effectively, there has to be a way of linking or ___________ them to meet the overall needs of an organism.

3. State one difference between anabolism and catabolism.

4. The following diagram shows two simple metabolic pathways.

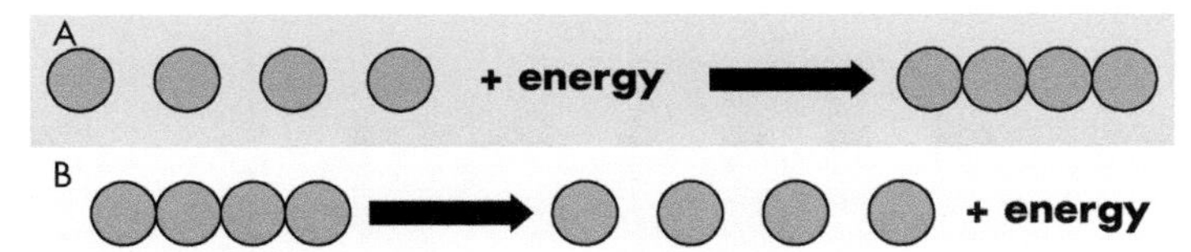

   Identify which pathway is anabolic and which is catabolic and give a reason for your answer.

> **Hint** Another command word. 'Identify' often means simply adding a label or a one-word answer.

5. Which row in the table is a correct description of a catabolic reaction?

	Substrate molecule	Product(s)	Energy
A	large	small	released
B	large	small	required
C	small	large	released
D	small	large	required

> **Hint** Remember that the energy currency molecule involved in metabolism is ATP.

6 State a property of some enzymes shown by the double-headed arrows in the following equation.

enzyme + substrate ⇌ enzyme-substrate complex ⇌ enzyme + product

> **Hint** Another command word. 'State' usually requires a one-word or very short answer.

7 When glucose enters a cell, it is modified irreversibly by an enzyme. Suggest an advantage to the cell of this enzymatic modification.

> **Hint** Another command word. 'Suggest' usually needs you to apply your knowledge to a new situation.

8 Red blood cells have an enzyme which can convert carbon dioxide, when it is high concentration, to a form which is easily transported in the plasma. This same enzyme can catalyse the reverse reaction to reform carbon dioxide.

Explain how this ability of the enzyme is useful in exchanging carbon dioxide.

9 The brain has a very high demand for energy and so brain cells often use a relatively short metabolic pathway, similar to what is shown here, to metabolise glucose.

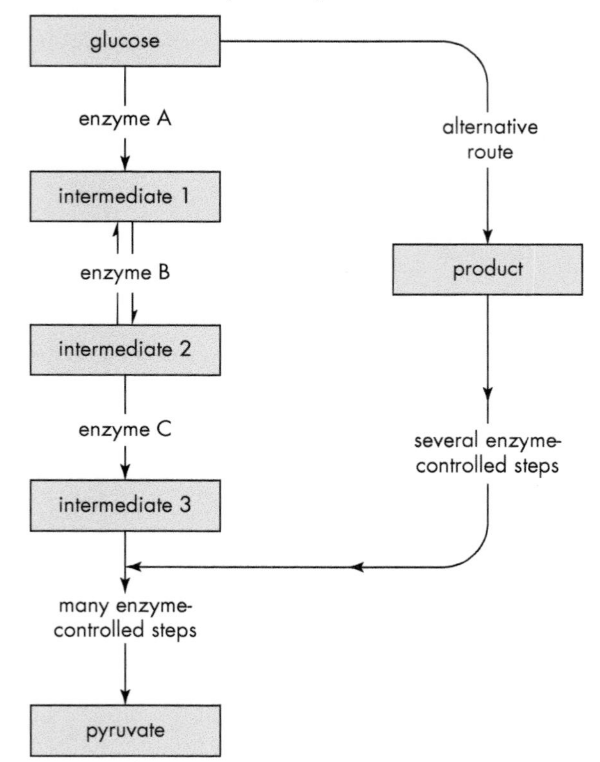

Explain why this type of alternative pathway is useful in brain cell metabolism.

> **Hint** Another command word. 'Explain' asks you to discuss why an action has been taken or an outcome reached. What are the reasons and/or processes behind the action or outcome?

10 The diagram below shows some of the features of the cell membrane.

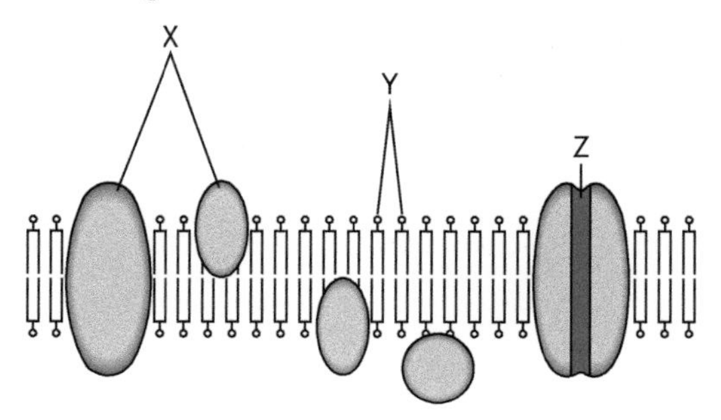

> **Hint** A good way to remember the structure and function of the cell membrane is to make a fully labelled diagram with the function of each part clearly indicated.

a Identify the molecules X, Y and Z on the diagram.

b State the function of molecule Z.

c Oxygen can diffuse easily across the cell membrane.

State which layer(s) of molecules the oxygen will pass through [use the letter].

d State what part of the cell membrane larger molecules, such as glucose, will pass through [use the letter].

11 An animal cell was found to have sodium ions in its surroundings at a concentration of 20 times greater than the concentration of sodium ions in its cytoplasm.

When the cell was cooled, the sodium ion concentration in the cytoplasm rose.

When the cell was warmed up to its normal temperature, the balance of the sodium ions was restored.

Which of the following explains the movements of the sodium ions?

	Inward movement	Outward movement
A	active transport	diffusion
B	diffusion	active transport
C	osmosis	active transport
D	active transport	osmosis

12 State one general function of an enzyme which is embedded in a cell membrane.

9B Control of metabolic pathways

1 Explain how the presence or absence of an enzyme can control a metabolic pathway.

2 Decide if each of the following statements relating to enzyme regulation in the table below is **TRUE** or **FALSE**. If the statement is **FALSE**, write the correct term to replace the term underlined in the statement.

Statement	True	False	Correction
Enzymes are coded for by genes.			
A substrate may not be synthesised if the specific enzyme is present.			

3. The diagram below shows one model to explain enzyme activity.

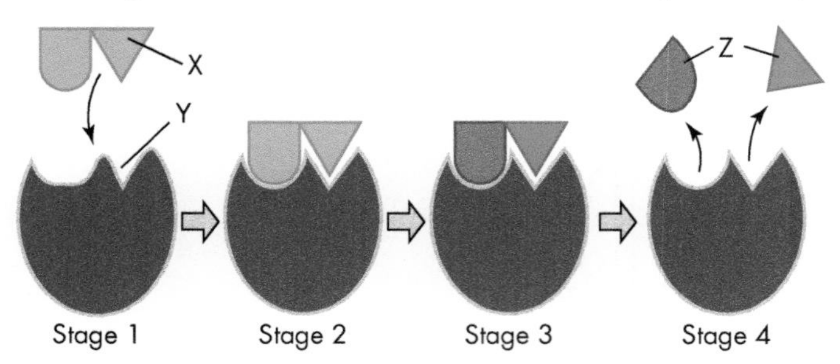

a State the name given to this model.

b Identify the substances labelled X, Y and Z.

c Explain what is happening at each of the four stages shown.

d State in which stage(s) there is/are a high and a low affinity between the reactants.

4. In order to start many enzyme-catalysed reactions, a small input of energy is required to make the substrate molecules react together.

a Give the term used to describe this energy.

b State the effect the presence of an enzyme has on this energy input.

5. The diagram below shows the energy changes in a reaction without and with an enzyme present.

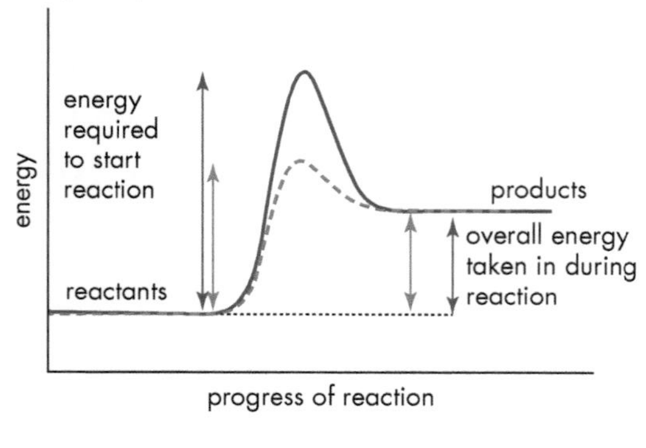

a Describe how the energy levels of the products compare with the energy levels of the reactants.

b Describe how the energy levels of the products when an enzyme is present compare with the energy levels of the products when an enzyme is absent.

6. Which of the following graphs shows the effect of increasing substrate concentration on the rate of an enzyme-catalysed reaction?

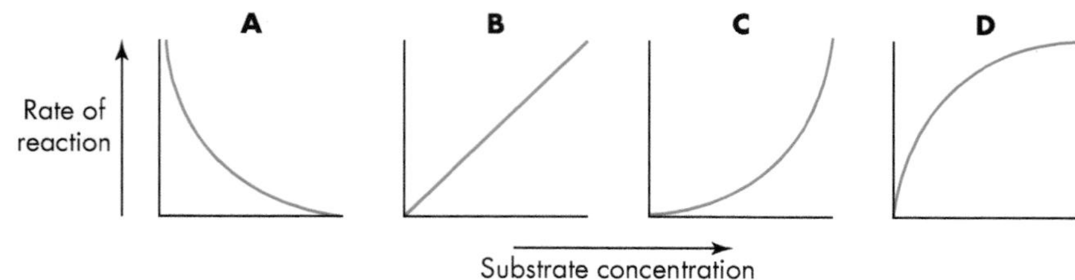

7 The following diagram shows an enzyme-catalysed reaction under three different conditions:

no inhibitor present

non-competitive inhibitor present

competitive inhibitor present.

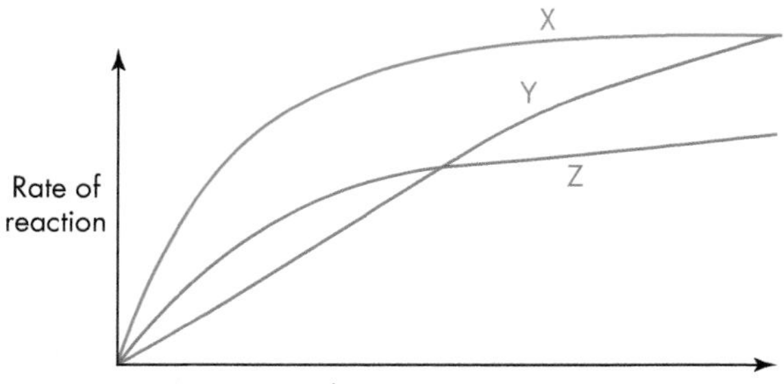

Which of the following best represents each curve?

	No inhibitor present	Non-competitive inhibitor present	Competitive inhibitor present
A	X	Z	Y
B	Z	Y	X
C	Y	X	Z
D	X	Y	Z

8 The following data were obtained for an enzyme-catalysed reaction in the presence and absence of an inhibitor.

Substrate concentration (units)	0.1	0.2	0.5	1.0	1.5
Rate of reaction with inhibitor (units product produced/min)	1.5	3.0	5.0	8.0	10.0
Rate of reaction without inhibitor (units product produced/min)	2.5	3.0	4.0	7.0	9.0

a Name the type of inhibition shown here.

b Explain your answer to **a**.

9C Short extended response question

Write notes on end-product inhibition as a mechanism for enzyme regulation.

Hint This is typical of a question usually worth four marks in the Higher Examination, so aim to give at least four points, which must be relevant to the question.

9D Longer extended response question

Write notes on the different types of metabolic pathways.

Hint This is typical of a question usually worth seven or more marks in the Higher Examination, so aim to give at least the name number of points, which must be relevant to the question.

9E Key terms

Link each key term below with the correct description.

	Key term		Description
1	metabolism	A	substance which binds irreversibly to an area, other than the active site of an enzyme, rendering the enzyme inactive
2	metabolic pathway	B	substance which can slow down or stop enzyme activity
3	catabolism	C	substance which binds reversibly to the active site of an enzyme reducing enzyme activity
4	anabolism	D	inhibition of an enzyme when product accumulates
5	inhibitor	E	change in shape of enzyme's active site when specific substrate fits
6	competitive inhibitor	F	position in space of reactants in an enzyme-catalysed reaction
7	non-competitive inhibitor	G	protein embedded in membrane which can actively transport molecules
8	phospholipid	H	protein which spans the membrane and allows large molecules to pass through its pore
9	protein pore	I	molecule of fat with phosphate group added
10	protein pump	J	synthesis of large molecules from small molecules
11	induced fit	K	breakdown of large molecules into small molecules
12	orientation	L	chain of biochemical reactions
13	end-product inhibition	M	all the chemical reactions which take place in a cell
14	affinity	N	organic catalyst
15	enzyme	O	degree of attraction between substrate and enzyme

9F Data handling and experimental design

1 The following diagram relates to the total number of substrate molecules that can be converted by a particular enzyme into products (unit volume/unit time) with different substrate concentrations (%).

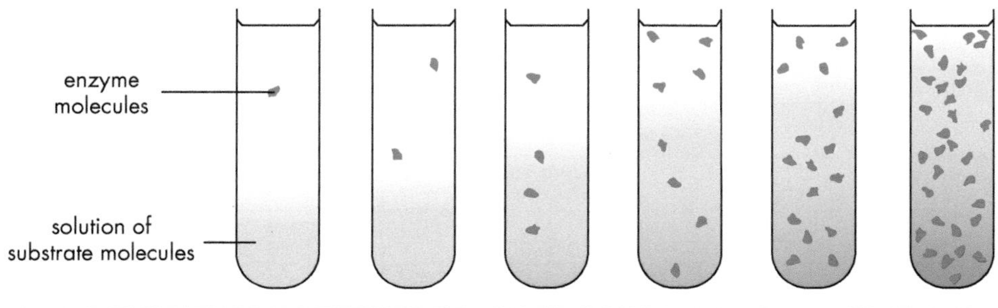

Tube	1	2	3	4	5	6
Substrate concentration (%)	2.5		10.0	20.0	40.0	80.0
Total number of substrate molecules converted into products (unit volume/unit time)	1	2	4	8		8

a Calculate the substrate concentration for tube 2.

b Predict the conversion of the substrate molecules into products at a substrate concentration of 40%.

> Hint Another command word, 'Predict' asks you to work out what will happen based on the information available.

c Identify evidence from the data which suggests that at a substrate concentration of 20%, some other factor is limiting the conversion rate.

d Identify the dependent variable here.

> Hint The dependent variable is the one which is actually measured and 'depends' on the independent variable, which is fixed by the experimenter.

2 The effect of increasing the concentration of an inhibitor (mol/l) on the activity of an enzyme (% control) is shown in the table below.

Concentration of inhibitor (mol/l)	Enzyme activity (% control)
0.0	100
0.1	90
0.2	75
0.3	30
0.4	10
0.5	0

Which of the following is a valid conclusion from these data? Inhibition was:

A lowest at high concentrations of the inhibitor
B highest at low concentrations of the inhibitor
C highest at high concentrations of the inhibitor
D lowest at a concentration of 0.2 mol/l.

3 A particular type of bacterial cell can manufacture an enzyme called protease. In the process, the bacterial cells use up glucose and produce dry mass material.

The graph below shows changes in the glucose concentration (g/100 cm^3) dry mass of the bacterial cells (g/l) and protease concentration (units/cm^3).

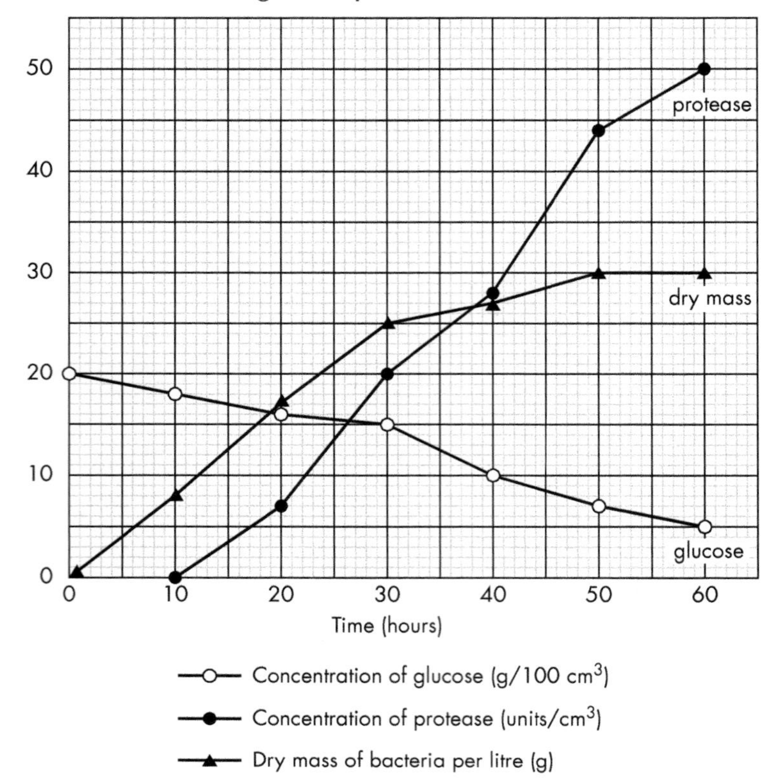

a Calculate the average increase per hour in the dry mass of the bacterial cells over the first 50 hours.

b Calculate the glucose consumed after 60 hours.

c State which 10-hour period produced the most protease.

d State how many hours passed before 25% of the glucose was used up.

10 Cellular respiration

10A Breakdown of glucose

Using the word bank provided, copy and complete the following description of the breakdown of glucose.

A word may be used once, more than once or not at all.

dehydrogenase – hydrogen – ATP – oxygen – water – ADP – electrons NAD – carrier – pyruvate – acetyl – glucose

When ___________ is broken down to carbon dioxide and ___________ in the presence of ___________, ___________ is produced as ___________ ions and electrons are removed by the action of ___________ enzymes. These ions and electrons are picked up by a ___________ molecule, such as ___________.

10B Role of ATP

1 The following diagram shows an important process which takes place during cell respiration to produce ATP.

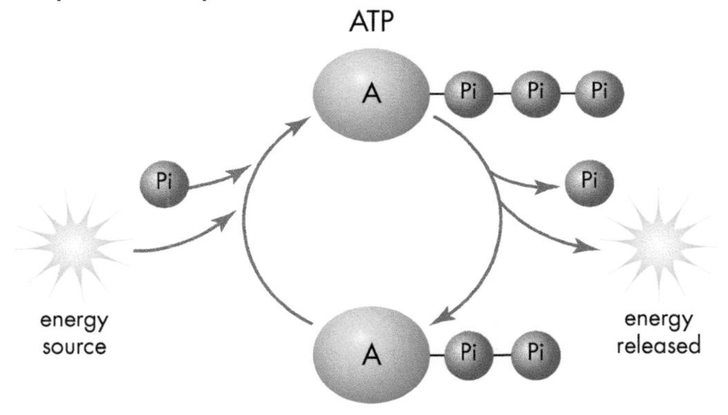

a State one direct source of energy to drive the reaction shown.

b State two cellular processes the energy released could be used for.

State the term used to describe the process shown in the diagram below.

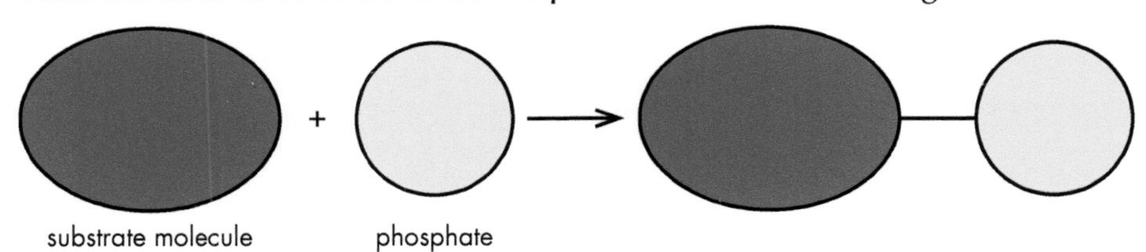

10C Metabolic pathways of cellular respiration

1 Which line in the table below shows where ATP is synthesised in a muscle cell respiring in the absence of oxygen?

	Cytoplasm	Mitochondrion
A	yes	no
B	yes	yes
C	no	no
D	no	yes

2 A piece of muscle was cut into three strands, X, Y and Z, each 60mm long.

Each strand was treated as shown in the table below.

The length of each strand was measured 15 minutes after the start of the experiment.

Muscle strand	Solution added to muscle strand	Length of muscle (mm)	
		Start	After 15 minutes
X	1% glucose	60	60
Y	1% ATP	60	40
Z	1% ATP boiled and cooled	60	58

Which of the following is a valid conclusion from these results?

A Muscles do not use ATP as an energy source.

B ATP is an enzyme.

C Muscles synthesise ATP in the absence of glucose.

D ATP is not an enzyme.

3 The table below contains information about respiration in a liver cell.

Information	Letter
Takes place in cytoplasm	A
Hydrogen ions and electrons released	B
Carbon dioxide released	C
Occurs in the absence of oxygen	D

Using the appropriate letter from the table, match the information to the stages of respiration shown below. One box has been completed for you.

Each letter may be used once, more than once or not at all.

Stage of respiration	Letter(s)		
Glycolysis			
Citric acid cycle	B		
Conversion of pyruvate to lactate			

4 Using the word bank provided, copy and complete the following description of some of processes involved in glycolysis.

A word may be used once, more than once or not at all.

hydrogen – oxygen – water – two – investment – pay-off – ATP lactate – pyruvate – phosphorylated – carrier – intermediate – four

Glycolysis does not require __________. The initial reactions of glycolysis use up ______ molecules of ATP as an energy __________ phase to provide energy to convert glucose to intermediate __________ compounds. This is followed by reactions which convert phosphorylated __________ compounds into two __________ molecules, which is an energy __________ phase because __________ ATP molecules are synthesised.

5 The diagram below shows a mitochondrion within the cytoplasm of a cell.

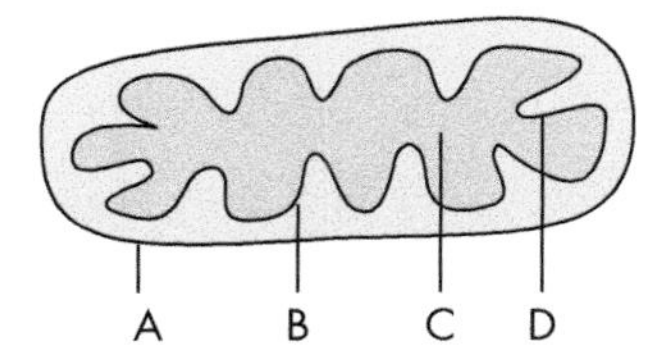

Identify where the regeneration of oxaloacetate takes place.

6 The diagram below shows some of the stages of aerobic respiration in a human liver cell.

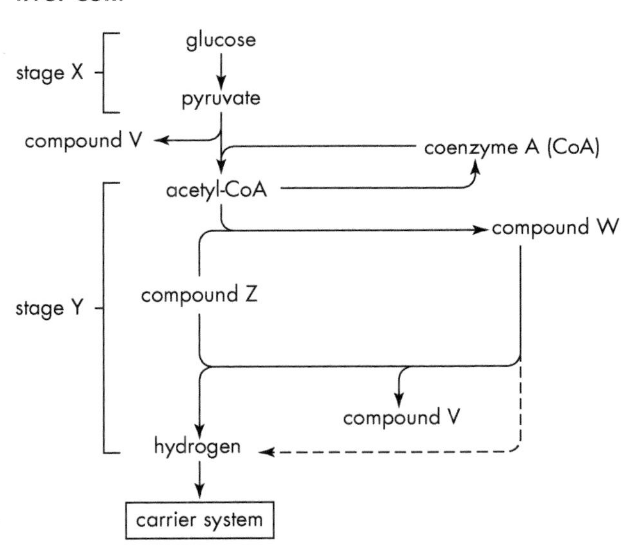

a Names stage X and Y.

b Name compounds V, W and Z.

c Describe the role of dehydrogenases found, for example, in stages X and Y.

d State the exact location of stage Y.

e Identify the carrier system shown.

f Describe what happens to pyruvate in this cell in the absence of oxygen.

7. State an important coenzyme present in the citric acid cycle that picks up hydrogen ions and electrons.

8. State what happens to pyruvate in a yeast cell when oxygen is unavailable.

9. Decide if each of the following statements relating to the electron transport system in the table below is **TRUE** or **FALSE**. If the statement is **FALSE**, write the correct term to replace the term underlined in the statement.

Statement	True	False	Correction
NAD is located on the outer membrane of the mitochondria.			
Within the electron transport system, in the presence of oxygen, NAD picks up hydrogen ions and electrons and transfers them to oxygen to form water.			

10D ATP synthesis

1. Using the word bank provided, copy and complete the following description of ATP synthesis.

   A word may be used once, more than once or not at all.

   hydrogen – oxygen – chain – mitochondria – aerobic – pay-off – protease
   matrix – ions – rotate – carrier – synthase – ATP – molecules – invert

   Energy, released as hydrogen __________ and electrons move down the electron transport __________, is used to pump the hydrogen ions from the __________ to the space between the two membranes of the __________. The return flow of these ions synthesises __________ by the action of the enzyme ATP __________ which is caused to __________. Almost all of the ATP synthesised during __________ respiration is synthesised in this way.

2. ATP synthase:

   A sits in the outer membrane of the mitochondrion

   B catalyses the breakdown of ATP

   C uses the flow of hydrogen ions across the mitochondrion

   D does not catalyse a phosphorylation reaction.

10E Short extended response question

Write notes on ATP synthase.

10F Longer extended response question

Write notes on glycolysis.

10G Key terms

Link each key term below with the correct description.

	Key term		Description
1	substrate	**A**	carrier molecule that accepts hydrogen ions
2	electron transport chain	**B**	enzyme which catalyses the removal of hydrogen from a substrate
3	phosphorylation	**C**	breakdown of glucose in the absence of oxygen
4	ATP synthase	**D**	molecule formed from glucose after glycolysis
5	pyruvate	**E**	series of reactions which occur in the matrix of a mitochondrion
6	citric acid cycle	**F**	membrane-bound enzyme which catalyses the synthesis of ATP
7	fermentation	**G**	addition of a phosphate group to a molecule
8	glycolysis	**H**	compound formed as an end-product of respiration without oxygen in an animal cell
9	oxaloacetate	**I**	initial series of reactions in respiration which occur in the cytoplasm
10	acetyl coenzyme A	**J**	alcohol produced by yeasts during fermentation
11	ethanol	**K**	intermediate compound in the citric acid cycle
12	NAD	**L**	intermediate compound linking glycolysis to the citric acid cycle
13	dehydrogenase	**M**	molecule on which an enzyme acts
14	citrate	**N**	series of reactions which occur on the inner membrane of a mitochondrion resulting in formation of water
15	lactate	**O**	intermediate compound in the citric acid cycle which joins with acetyl coenzyme A to form citric acid

10H Data handling and experimental design

Glycogen is the primary storage carbohydrate in muscle cells which can be converted into glucose for respiration.

Endurance time is how long a muscle can work without tiring.

The table below shows data for the average endurance time (min) for volunteers who had different concentrations of glycogen (g/kg muscle) in their muscles at the start of an exercise.

Average endurance time (min)	50	60	100	125	150	160	175
Initial muscle glycogen concentration (g/kg muscle)	5	10	15	20	25	30	35

a Plot a line graph of these data.

b Calculate the difference in the initial muscle glycogen concentration between the group whose average endurance time was 50 minutes compared to those whose average endurance time was 100 minutes.

c It was found in another group of volunteers that even when they had an initial muscle glycogen concentration of 0 g/kg muscle, their average endurance time was 30 minutes.

Suggest why the muscles of these volunteers were still able to function.

d Calculate the percentage increase in the average endurance time when the initial muscle glycogen rises from 5 to 10 g/kg muscle.

e State the simplest whole number ratio of the average endurance time for volunteers who had 10, 20 and 30 g/kg muscle at the start of the investigation.

f Identify the dependent variable here.

g State an aspect of the experimental design which could lead to inaccurate results.

h State one experimental design feature which could lead to unreliable results.

11 Metabolic rate

11A Measurement of metabolic rate

1. Decide if each of the following statements relating to measuring metabolic rate in the table below is **TRUE** or **FALSE**. If the statement is **FALSE**, write the correct term to replace the term underlined in the statement.

Statement	True	False	Correction
Metabolic rate is rate at which energy is released in a fixed time.			
The rate of respiration can be measured by how much carbon dioxide is produced in a given time.			

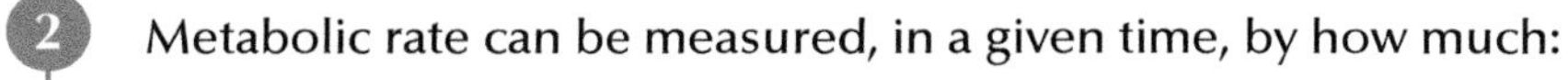

2. Metabolic rate can be measured, in a given time, by how much:

A carbon dioxide is produced and oxygen used up

B heat energy is released and oxygen produced

C oxygen is produced and carbon dioxide used up

D heat energy is released and carbon dioxide used up.

11B Transporting oxygen to cells

1. Fish have a heart with:

A one ventricle and one atrium

B two atria and one ventricle

C two ventricles and one atrium

D two ventricles and two atria.

2. Identify which vertebrates are associated with each of the circulatory systems shown.

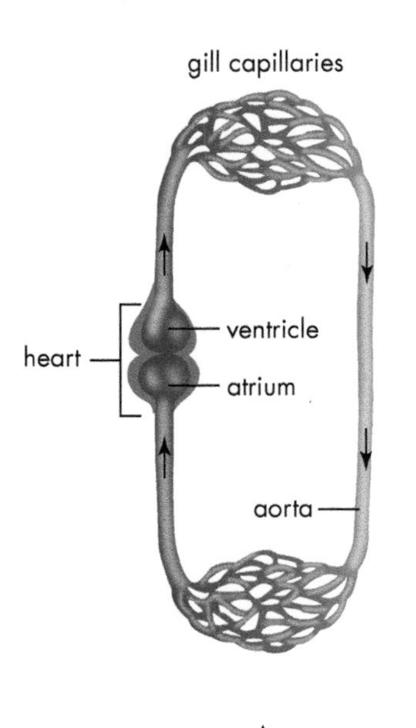

A

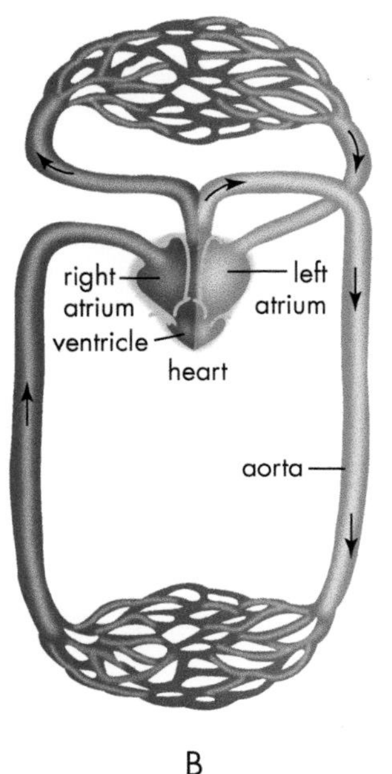

B

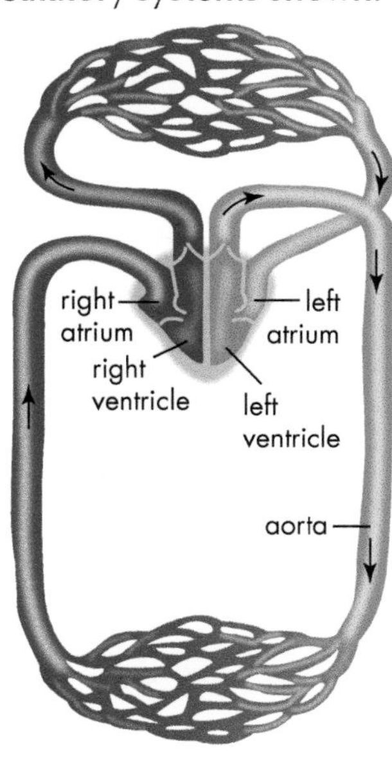

C

3 Using the word bank provided, copy and complete the following description of different vertebrate hearts.

A word may be used once, more than once or not at all.

one – reptiles – heart – once – ventricles – twice – mammals – fish – pressure
atria – low – circuit – high – double – gills – ventricle

Fish have a heart with ____ atrium and ____ ventricle. The blood flows in ____ direction only, from the heart to the ____ and back to the ____ again. Amphibians and ______ have a heart with one ________ and two atria. The blood flows through the heart ______ for each ________ of the body. The ______ circuit of the blood helps to maintain ________ inside the blood vessels.

In __________ and birds, the heart is fully divided into two upper ______ and two lower ____________. The blood flows through the heart twice for each circuit of the body. Pressure of the blood is kept ______ throughout.

11C Short extended response question

Describe how resting metabolic rate in mammals changes as the body mass increases.

11D Longer extended response question

Give an account of the transport of oxygen in amphibians and reptiles.

11E Data handling and experimental design

The volumes of air breathed in and out of the lungs, as well as the capacities of lungs, can be measured and plotted as a graph. These are shown below for one girl.

Each peak shows a volume of air, measured in cm^3, breathed in and each trough shows a volume of air breathed out. Small fluctuations show quiet breathing while large fluctuations show maximum possible volumes which can be breathed in and out.

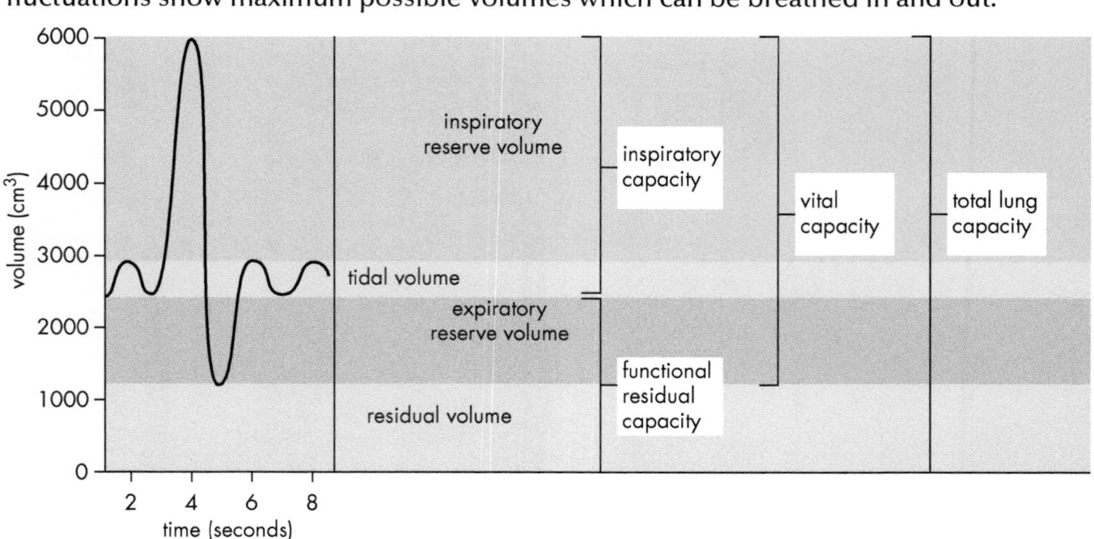

a The tidal volume is the volume of air which can be breathed out or breathed in at rest.

Calculate the tidal volume for this girl.

b Calculate how many complete breaths/minute this girl takes at rest.

c The functional residual capacity is the total volume of the expiratory reserve volume and residual volume.

Calculate the functional residual capacity for this girl.

d The vital capacity is the total volume of the inspiratory capacity and the functional residual capacity.

A boy has a vital capacity of 5000 cm^3 and a functional residual capacity of 2500 cm^3.

Calculate the inspiratory capacity for this boy.

e The total lung capacity is the total volume of air a lung can hold.

State the total lung capacity for this girl.

f State the time when the girl started to yawn.

g State one aspect of the design of this experiment which could give rise to unreliable results.

12 Metabolism in conformers and regulators

12A Effect of external factors

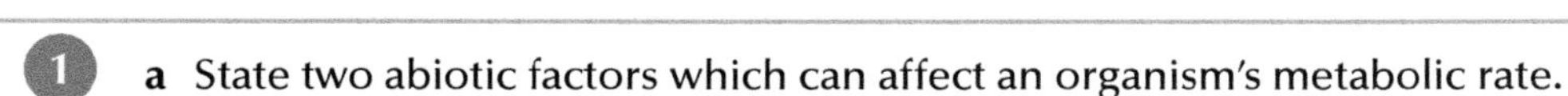

1 **a** State two abiotic factors which can affect an organism's metabolic rate.

b Explain how one of these abiotic factors can affect an organism's metabolic rate.

2 Which of the following can both affect the metabolic rate of an animal?

A Light intensity and temperature

B Temperature and pH

C Salinity and light intensity

D Temperature and light intensity

12B Conformers and regulators

1 Explain the difference between a conformer and a regulator.

2 Which of the following is **not** a conformer?

A Reptile

B Mammal

C Amphibia

D Insect

3 Decide if each of the following statements relating to gas conformers and regulators in the table below is **TRUE** or **FALSE**. If the statement is **FALSE**, write the correct term to replace the term underlined in the statement.

Statement	True	False	Correction
Conformers must adapt their behaviour by moving to environments which are less favourable.			
Conformers usually live in stable environments.			
Birds have an internal environment which is not dependent on the external environment.			
Regulators can live across a narrow range of environments.			

4. The following graph shows the relationship between the external temperature and the internal body temperature of two different animals, X and Y.

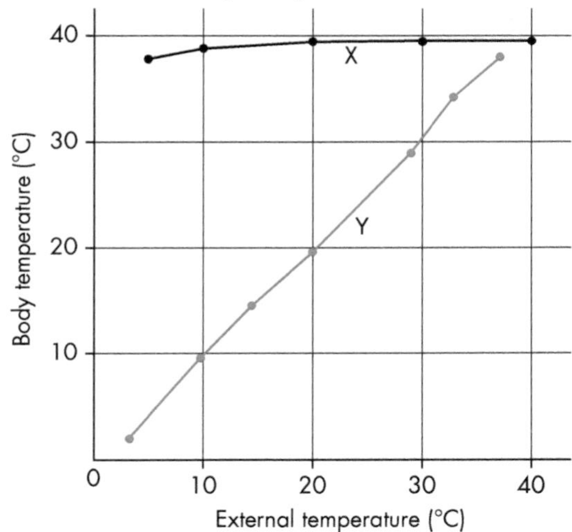

Which of the following correctly identifies the two different animals?

	Animal X	Animal Y
A	regulator	regulator
B	conformer	conformer
C	regulator	conformer
D	conformer	regulator

5. Describe the range of ecological niches occupied by conformers and regulators.

6. State what is meant by homeostasis.

7. Explain why regulators tend to have high metabolic rates.

12C Negative feedback control and thermoregulation in mammals

1. Using the word bank provided, copy and complete the following description of negative feedback.

A word may be used once, more than once or not at all.

broad – negative feedback – conformers – hypothalamus – brain narrow – effectors – regulators – nervous – receptors – temperature environment – set point – variables – skin – hormonal

Most physiological ____________, such as temperature, are controlled within very ____________ limits above or below a ____________. An important mechanism to control this is called ____________, common in ____________, which responds to a change in the ____________. As the ____________ of a mammal's body changes, this is detected by ____________ generating ____________ signals which go to the ____________ in the ____________. This sends out nervous signals to ____________ which respond by activating structures in the ____________ and muscles. The response attempts to restore the rise above or fall below the ____________ back to normal.

2 Homeostasis acts:

A to maintain physiological variables above or below the set point

B to keep a relatively constant internal environment

C to change the set point for a physiological variable

D usually by positive feedback.

3 The following diagram shows a generalised negative feedback mechanism.

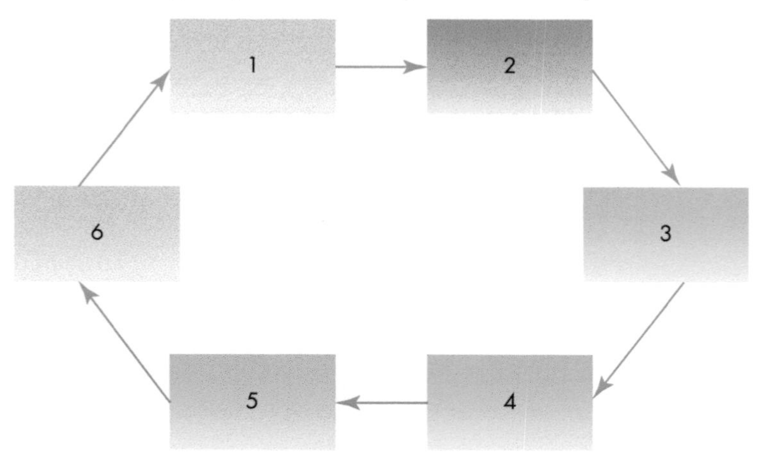

Hint Flow diagrams, such as this one, appear very often in biology. They are an excellent tool for revision and consolidating information which is sequential in nature.

Copy and complete the diagram by linking the correct term(s) below with the numbered blank boxes:

- set point
- change above or below set point
- receptors detect change above or below set point
- return to set point
- effectors react to reverse the change
- change is communicated to effectors

4 Which of the following correctly shows information about how a mammal regulates its body temperature?

	Method of signalling	Effector organ	Receptor site
A	hormonal	muscle	hypothalamus
B	nervous	liver	skin
C	hormonal	brain	hypothalamus
D	nervous	skin	hypothalamus

5. Which of the following would result if skin was exposed to high temperature?

	Hair erector muscle	Change in blood flow to capillaries in skin
A	relaxed	increased
B	contracted	decreased
C	relaxed	decreased
D	contracted	increased

6. State the part of the brain which regulates body temperature.

12D Short extended response question

Discuss the importance of regulating temperature in maintaining metabolism.

12E Longer extended response question

Discuss negative feedback in relation to thermoregulation in mammals.

12F Key terms

Hint Flash cards made from these key terms are powerful learning devices.

Link each key term below with the correct description.

1	conformer	A	general term for maintenance of body systems
2	regulator	B	value for some variable that is kept at a constant level
3	homeostasis	C	specialised group of cells that can detect change
4	receptor	D	region of brain that can monitor temperature
5	hypothalamus	E	specialised group of cells that can react to change
6	effector	F	organism that cannot maintain its internal metabolic rate in a changing environment
7	set point	G	organism that can maintain its internal metabolic rate in a changing environment

12G Data handling and experimental design

An investigation was carried out to look at the effect of exercise on the production of sweat.

A treadmill was placed in a room with a constant temperature.

A 25-year-old female athlete used the treadmill for six different lengths of time (measured in seconds) and the average rate of sweat produced (measured as mg/cm^2 skin/minute) recorded as shown in the table below.

After each exercise period, a rest of 5 minutes was given.

Exercise trial	Length of exercise period (s)	Average rate of sweat produced (mg/cm^2 skin/minute)
1	20	0.05
2	40	0.10
3	60	0.25
4	80	0.30
5	100	0.35
6	120	0.40

a Draw a line graph to show the length of the exercise period against the average rate of sweat production.

Hint Make sure you draw line graphs point to point with a ruler.

b State three variables that should be keep constant to allow a valid comparison of the exercise trials.

c State how the reliability of the results could be improved.

d Explain why a rest period was given after each exercise period.

e Calculate the total sweat produced during exercise trial 6.

f Predict the average rate of sweat production during an exercise period of 140 seconds.

g What is the dependent variable here?

Hint Remember the dependent variable is the one being measured which depends on the independent variable.

13 Metabolism and adverse conditions

13A Surviving adverse conditions

1 Using the word bank provided, copy and complete the following paragraph which relates to surviving adverse conditions.

A word may be used once, more than once or not at all.

limits – temperature – adaptations – homeostasis – habitat
metabolic – slow down – tolerate – speed up – fatal – conserve

An organism can only survive in its ___________ if the ___________, water and food availabilities stay within certain ___________. If these conditions go beyond the limits which the organism can ___________, its metabolism will start to ___________ ___________ with potentially ___________ consequences. Some organisms can ___________ energy by slowing down their ___________ rate. Many such organisms have evolved ___________ to maintain ___________ or avoid the harsh conditions altogether.

2 In hot summer conditions, snails retreat into their shells until the weather conditions improve.

This is an example of:

A aestivation

B hibernation

C predictive dormancy

D daily torpor.

Hint: Make sure you get these terms sorted – they are often confused.

3 Which of the following is an example of consequential dormancy?

A Plants slowing down their metabolism as the daylength decreases

B Hedgehog slowing down its heartbeat as daily temperature falls

C Dormouse stocking up food reserves in the form of body fat in autumn

D Toad burying itself to avoid extreme drought conditions

4 State one advantage and one disadvantage of consequential dormancy.

5 Decide if each of the following statements relating to hibernation, aestivation and daily torpor in the table below is **TRUE** or **FALSE**. If the statement is **FALSE**, write the correct term to replace the term underlined in the statement.

Statement	True	False	Correction
In harsh winter conditions, aestivation allows animals to survive by reducing their heartbeat.			
When water becomes in short supply, some animals can burrow underground to undergo hibernation.			
A decreased rate of activity when food supplies are limited may produce a state of daily torpor.			

13B Avoiding adverse conditions

1. The arctic tern can fly huge distances as it migrates.

   State two advantages to the tern's survival by performing this feat.

2. Migration:

   A is always over long distances

   B only involves learned behaviour

   C only involves innate behaviour

   D avoids metabolic adversity.

3. Which of the following is **TRUE**?

   A Innate behaviour is not genetically based.

   B Migratory behaviour is a product of evolution.

   C Learned behaviour is largely genetically based.

   D Migration is rarely linked to seasonal changes.

   Hint: If you are not sure answering this type of question, try to eliminate those answers which you know are definitely false first.

4. Migratory birds kept in captivity during their normal migratory period show extreme restless behaviour.

   Complete the following sentence by underlining the correct term.

   The restless behaviour shown by these birds strongly suggests the influence of their [genes / environment].

13C Short extended response question

Compare predictive and consequential dormancies.

13D Longer extended response question

Give an account of migratory behaviour and two ways it can be tracked.

13E Key terms

Link each key term below with the correct description.

	Key term		Description
1	displacement experiment	A	organism goes into a state of dormancy after adverse conditions develop
2	innate behaviour	B	period of time in which an organism is in a state of low metabolism during adverse conditions
3	learned behaviour	C	state of inactivity and low metabolism in response to high temperatures
4	daily torpor	D	state of low metabolism and body temperature in organisms over winter
5	migration	E	movement of animals linked to seasonal changes
6	hibernation	F	short-lived state in which an animal conserves energy by lowering body temperature and metabolism
7	aestivation	G	investigating migration by relocating animals then releasing them from new location
8	adaptation	H	behaviour shaped by experience
9	dormancy	I	behaviour shaped by inherited genes
10	predictive dormancy	J	characteristic giving organism an increased chance of survival
11	consequential dormancy	K	organisms goes into a state of dormancy before adverse conditions develop

13F Data handling and experimental design

The following graph shows the optimum temperature for growth of five different groups of bacteria. The information which follows shows the range of temperatures within which they survive and reproduce.

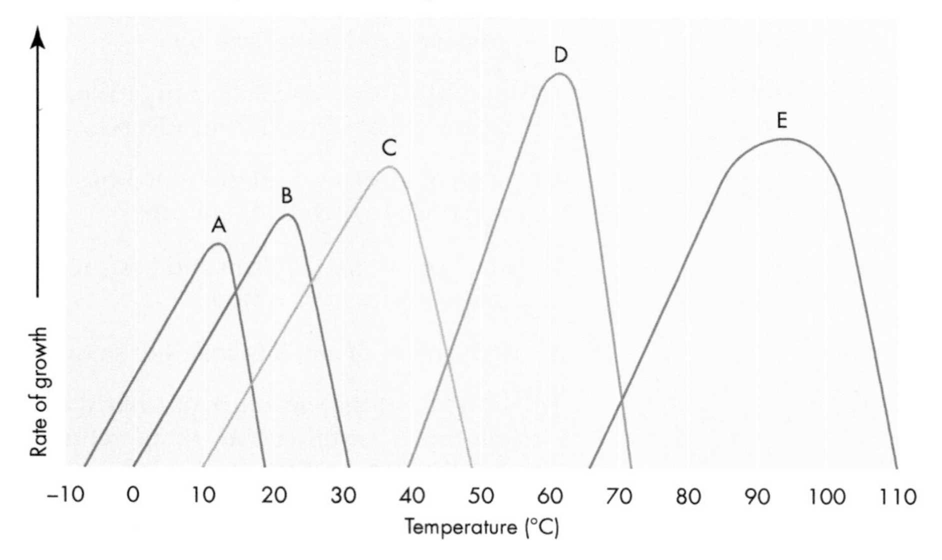

Group 1: –10–20 °C

Group 2: 10–50 °C

Group 3: 0–30 °C

Group 4: 65–110 °C

Group 5: 40–72 °C

a Match each group to the letters shown on the graph.

b State which group can grow over the widest range of temperature.

c State which group can grow over the narrowest range of temperature.

d State which group(s) could grow at a temperature of 45 °C.

e State which group(s) could not grow at a temperature of 68 °C.

2 During hibernation, a hedgehog can have periods when it wakes up and searches for food or water.

Graph 1 below shows the heartbeat (beats/minute) for one hedgehog over one of these periods, which lasted 10 hours.

Graph 1

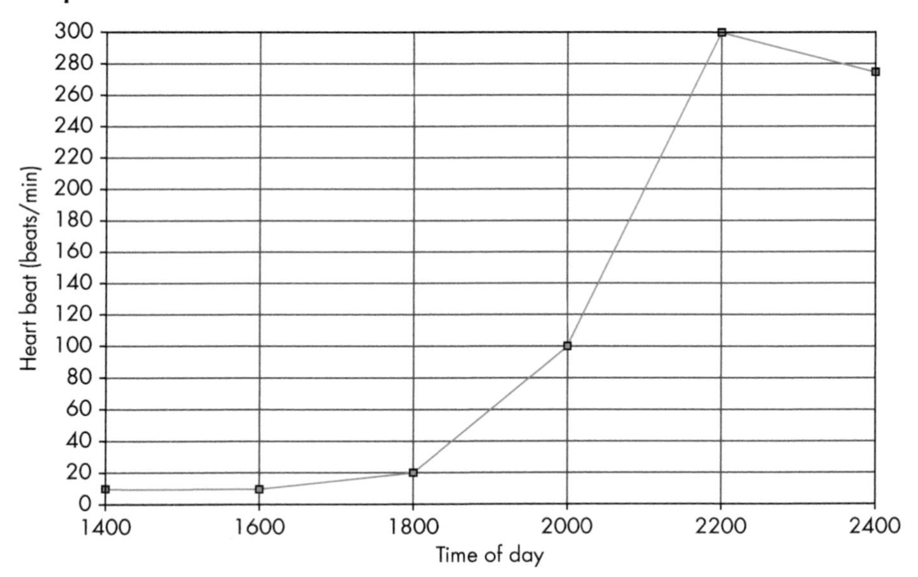

Graph 2 below shows the body temperature (°C) for the same hedgehog over the same 10-hour period.

Graph 2

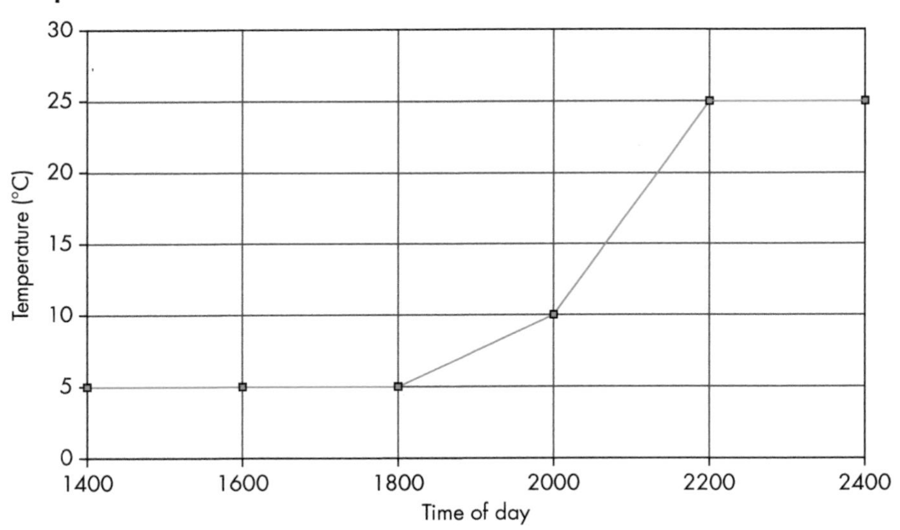

a Describe the changes in the heartbeat of the hedgehog from 1800 until 2200.

b Calculate the percentage increase in the heartbeat of the hedgehog from 1400 until 2000.

> Hint: To calculate percentage change, take the difference and divide it by the original then multiply by 100.

c State between which 2-hour period the greatest change in the body temperature of the hedgehog occurred.

14 Environmental control of metabolism

14A Growing microorganisms

1. Complete the following table describing some features of the three different domains.

Feature	Bacteria	Archaea	Eukaryota
Nucleus present in cells		No	Yes
Organelles present in cells	No		
Chromosome shape	Circular	Circular	
Cellular arrangement		Unicellular	Unicellular and multicellular

Hint Presenting and/or storing information in tabular form is a good way of remembering information.

2. Which of the following is not true?

A Microorganisms may be prokaryotic or eukaryotes.

B Prokaryotes include bacteria and archaea.

C Eukaryotes include algae, protozoa and fungi.

D Animal cells belong to the prokaryotes.

3. Using the word bank provided, copy and complete the following paragraph which relates to microorganisms.

A word may be used once, more than once or not at all.

industry – species – metabolism – variety – substrates – adaptable – niches enzymes – products – culture – manipulation

Microorganisms are very ___________ because they can grow on a ___________ of different ___________ for their ___________. Different ___________ of microorganisms are found in different ecological ___________ based on these different ___________. They are widely exploited in both ___________ and research because they grow very quickly, are ___________ and easy to ___________. Genetic ___________ of microorganisms can produce a wide variety of useful ___________.

4. State three environmental factors which can influence the growth of microorganisms.

5 Describe two ways in which energy may be supplied to microorganisms for biosynthesis.

6 Decide if each of the following statements relating to growing microorganisms in the table below is **TRUE** or **FALSE**. If the statement is **FALSE**, write the correct term to replace the term underlined in the statement.

Statement	True	False	Correction
Photosynthetic microorganisms use chemical energy directly for biosynthesis.			
Simple compounds such as amino acids can be used for microbial growth.			
Few microorganisms can produce all the complex molecules required for growth.			

7 Identify the two different types of growth media shown below.

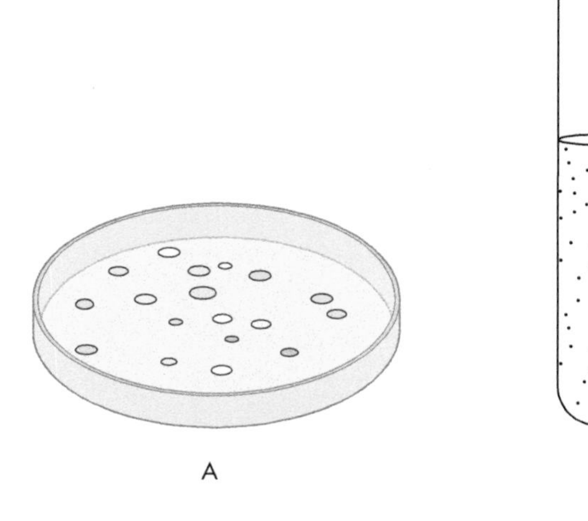

8 Some microorganisms cannot produce all the complex molecules they need for growth.

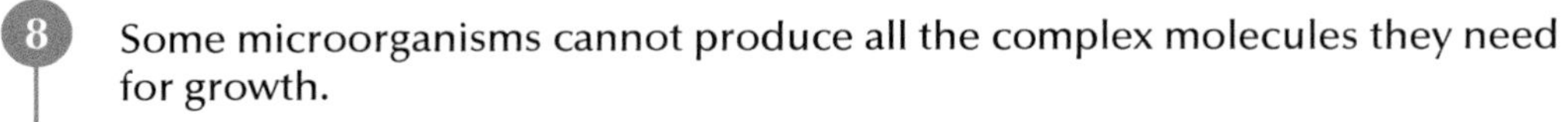

State two additional complex molecules which have to be added to the growth media to encourage these microorganisms to grow.

9 Explain why culturing microorganisms requires sterile conditions.

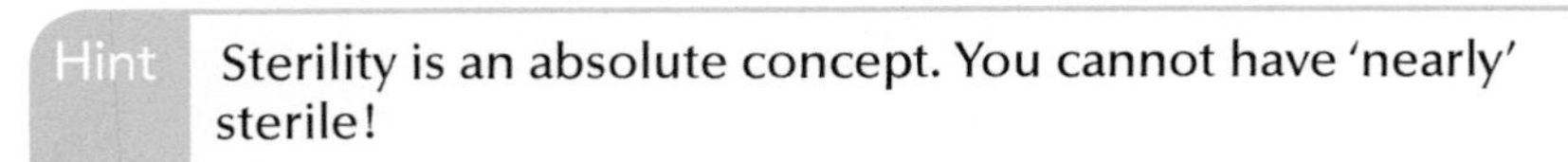

Hint Sterility is an absolute concept. You cannot have 'nearly' sterile!

10 Explain the role of a buffer in the media used to growth microorganisms.

11 Which of the following need to be controlled to grow bacteria found on the human skin in the lab?

A Light intensity, pH and oxygen level

B Oxygen level, temperature and pH

C Nutrient supply, light intensity and temperature

D Temperature, pH and light intensity

12 The following apparatus allows microorganisms to be grown on a large scale.

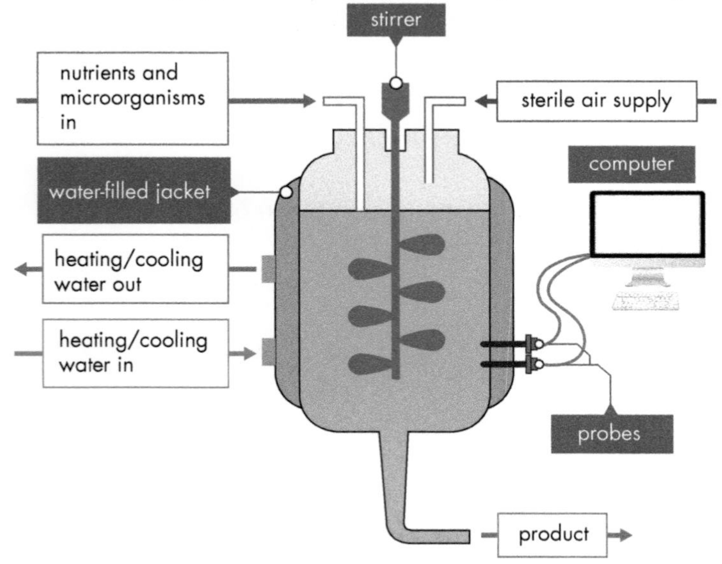

a State two variables the probes will feed back to the computer.

> **Hint** Consider the factors which might affect the enzymes present in the microbial cells.

b Suggest the function of the water-filled jacket.

c Suggest the function of the stirrer.

d Explain why the air supply needs to be sterile.

14B Phases of microbial growth

1 The graph below shows the change in the number of microbial cells against time.

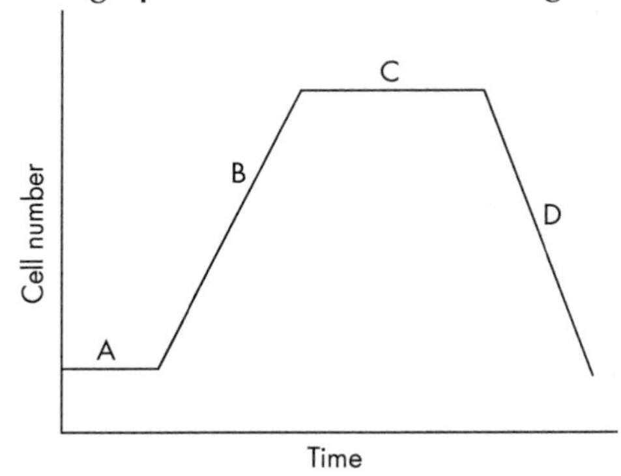

a Name each of the four phases, A, B, C and D, labelled on the graph.

b Complete the following table of the phases shown on the graph.

Phase	Description
A	high rate of metabolism and new enzymes being produced
B	number of microorganisms increases exponentially
C	
D	

c In culturing microorganisms it is important to control pH levels in the culture medium.

> **Hint** Make sure you know other variables which can affect microbial growth.

Describe how the pH of a culture medium can be controlled.

d Complete the following sentence by inserting a suitable term in the space provided.

The time it takes for one cell to divide into two is called the doubling time or ___________ time.

2 A viable count in a given volume of a culture measures:

a all the living and dead cells present

b only the living cells present

c only the dead cells present

d the total count minus the number of living cells present.

3 A total count in a given volume of a culture measures:

a all the living and dead cells present

b only the living cells present

c only the dead cells present

d all the cells present minus the viable count.

4 A certain blue dye only penetrates dead microbial cells.

Explain how this could be used to obtain a viable count of microorganisms in a culture.

14C Short extended response question

Give an account of the conditions required to culture bacteria in the lab.

14D Longer extended response question

Name and describe the different phases of microbial growth.

14E Key terms

Link each key term below with the correct description.

	Key term		Description
1	fermenter	**A**	measure of all cells in a culture
2	generation time	**B**	measure of only living cells in a culture
3	log phase	**C**	in growth the doubling of cell numbers every cycle
4	total count	**D**	time taken for one cell to become two cells
5	stationary phase	**E**	period when microorganisms are adapting to new growth environment
6	doubling time	**F**	another name for generation time
7	exponential	**G**	period when microorganisms are dying
8	lag phase	**H**	metabolism which occurs in the stationary phase of microbial growth
9	viable count	**I**	period when cell numbers double for each cycle
10	sterile	**J**	period when growth and death rates balance each other
11	aseptic technique	**K**	large container used to grow microorganisms on commercial scale
12	death phase	**L**	procedure used to prevent contamination of culture or environment
13	secondary metabolism	**M**	absence of all living material

14F Data handling and experimental design

A student carried out the following procedure.

A nutrient agar plate had 1 cm^3 of a bacterial culture added. The culture was then subject to a series of dilutions into sterile broth as shown below.

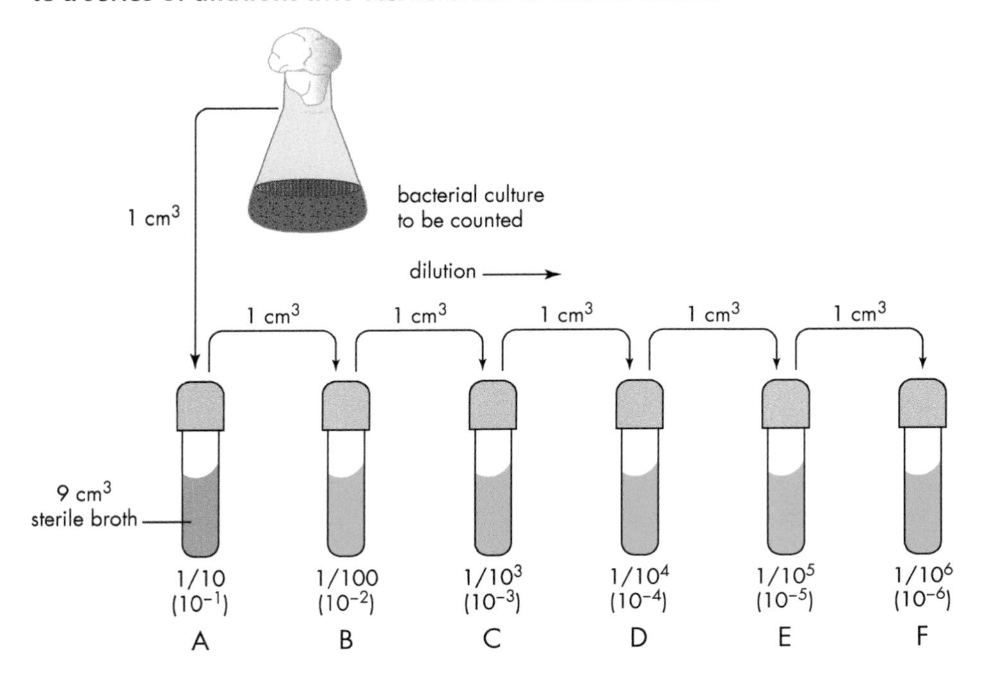

From each dilution, 1 cm^3 was plated out onto sterile nutrient agar plates and then incubated as shown below.

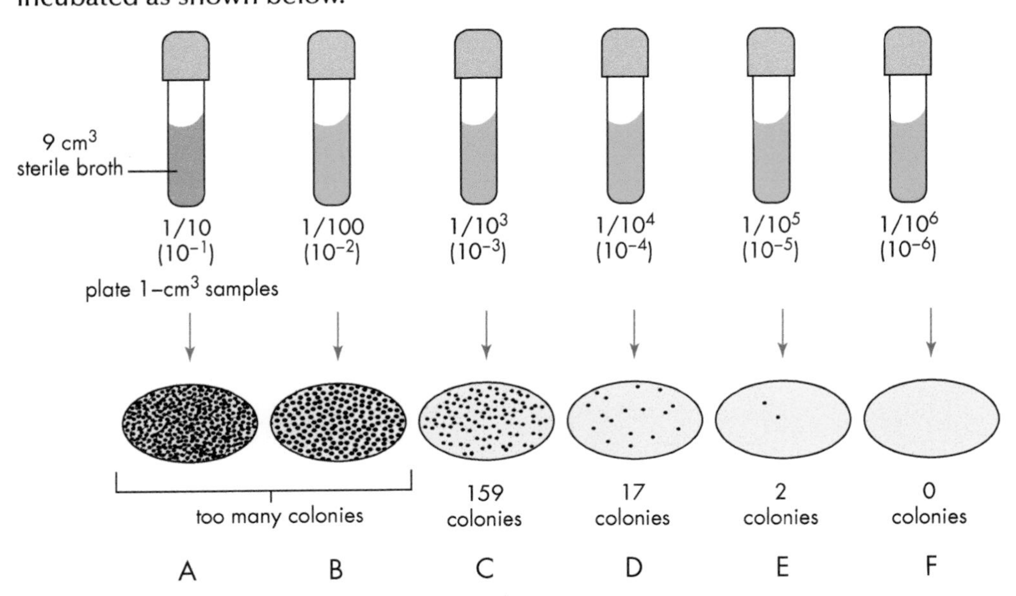

Each colony represents one bacterial cell in the original sample to be counted.

a Using the results from plate C, estimate the number of cells in the original sample to be counted.

b Identify which plate would give an estimated number of cells in the original sample as 200 000.

c State how the procedure could be improved to obtain more reliable results.

2 A bacterium undergoes six complete divisions in 2 hours.

Calculate how long it takes for one division to take place.

3 Calculate how many complete divisions it would take for three bacterial cells to produce 48 cells.

15 Genetic control of metabolism

15A Improving wild strains of microorganisms

1. Using the word bank provided, copy and complete the following paragraph, which relates to improving wild strains of microorganisms.

   A word may be used once, more than once or not at all.

ethanol – viruses – useful – synthesise – sugar – strains – yeasts – map toxic – genetically – biofuel – modification

   Scientists have found __________ of microorganisms which __________ products which are __________ to humans. For example, __________ produce __________ from __________ but now it is possible to __________ the genes responsible for this conversion so that it might be possible to use __________ modified yeasts to make __________. While __________ is __________ to growing yeast cells, genetic modification can be used to engineer them to tolerate high levels of __________ making them even better at synthesising biofuel.

2. Decide if each of the following statements relating to genetically improving microorganisms in the table below is **TRUE** or **FALSE**. If the statement is **FALSE**, write the correct term to replace the term underlined in the statement.

Statement	True	False	Correction
Agents which damage DNA are termed <u>mutagenic</u>.			
Mutant strains of microorganisms <u>cannot</u> back-mutate to their original genetic makeup.			

3. Name two techniques used to genetically improve microorganisms.

4. Which of the following is not likely to be an improvement to a strain of microorganism?

   A Growth which is rapid

   B Genetically stable cells

   C Non-toxic to humans

   D Decreased productivity

5. Suggest why producing large cell sizes is a target for improving microorganisms.

6. Which of the following is **TRUE**?

   A Yeasts can only reproduce sexually.

   B Sexual reproduction gives rise to variation.

   C Bacterial reproduction most commonly gives rise to genetic variation.

   D A mutated microorganism is always inferior to the wild strain.

1. Decide if each of the following statements relating to recombinant DNA technology in the table below is **TRUE** or **FALSE**. If the statement is **FALSE**, write the correct term to replace the term underlined in the statement.

Statement	True	False	Correction
Recombinant DNA technology allows the transfer of genetic material between different species.			
Transferring genetic material from a human cell into a bacterial cell often involves a vector.			

2. Give three examples of vectors used in recombinant DNA technology.

3. The following diagram shows a process involved in recombinant DNA technology.

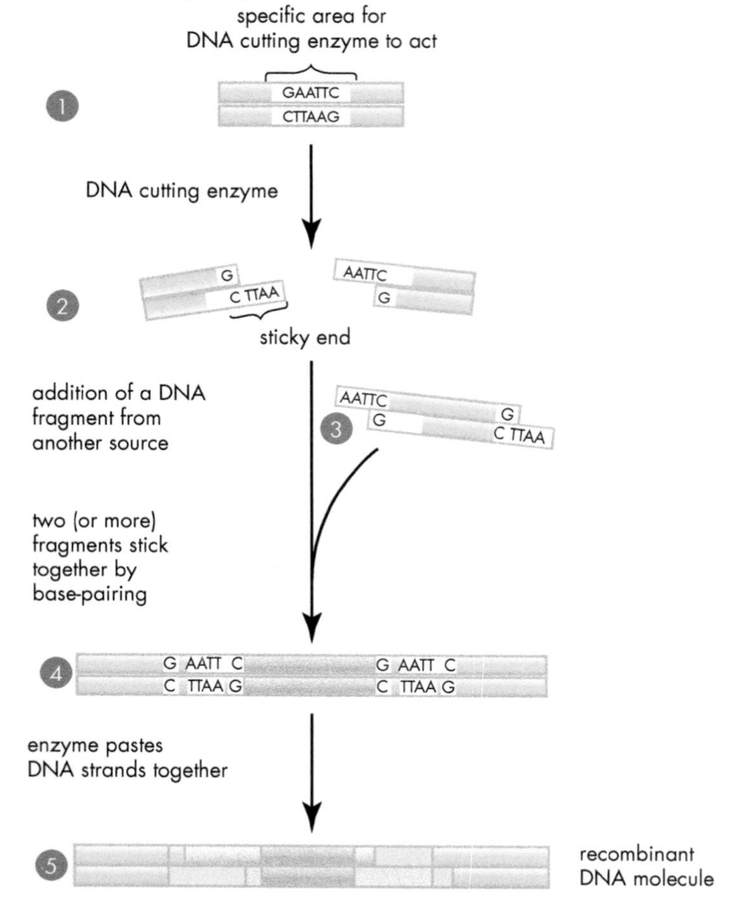

a Name the specific area of the DNA being cut.

b Identify the two enzymes shown.

c Suggest why the cut ends of DNA are referred to as 'sticky'.

4. A bacterial plasmid was used to transfer a desired gene into a plant.

   In order to identify which plant cells had taken up the desired gene, another gene, which confers antibiotic resistance, was also transferred.

   a Explain how the plants with the desired gene can be selected for.

   b Name a gene, such as the one conferring antibiotic resistance, which is used to identify genetically modified cells.

5. Using the word bank provided, copy and complete the following paragraph, which relates to restriction enzymes.

   A word may be used once, more than once or not at all.

similar – double – restriction – ligase – single – different – easy – restriction site sticky end – difficult – desired – two – site – end – plasmid – complements

   A ___________ enzyme does not cut across a ____________ strand of DNA directly. This would produce many ___________ pieces of DNA and make it ___________ to cut out an entire _________ gene. A _________ enzyme cuts across a _________ strand of DNA at _________ different points. The actual area where the enzyme performs this cut is called a ___________ ______ and results in the formation of a _______ _______. The same ___________ enzyme makes a similar cut in a bacterial _________ so that the desired gene's base sequence on the ______ ______ ___________exactly that of the ___________.

6. Suggest one advantage of using a chromosome made artificially in the lab as a vector in recombinant DNA technology.

15C Short extended response question

Write notes on the use of mutagenesis to improve wild strains of microorganisms.

15D Longer extended response question

Describe how recombinant DNA technology can be used to produce a desirable human product, such as the hormone insulin.

15E Key terms

Link each key term below with the correct description.

	Key term		Description
1	origin of replication	A	plasmid, virus or piece of artificial chromosome which carries genes to other cells
2	marker gene	B	particular part of DNA where replication takes place
3	vector	C	section of DNA which allows a transformed host cell to be identified
4	restriction endonuclease	D	enzyme which joins two detached strands of DNA
5	restriction site	E	enzyme which can split DNA at a specific position
6	ligase	F	short sequence found on both strands of a DNA molecule which can be recognised by a restriction enzyme
7	mutagenesis	G	process where genetic information of an organism is altered producing a mutation
8	recombinant DNA	H	end of DNA where one strand continues beyond the other by a few bases
9	artificial chromosome	I	yeast cell whose DNA has been altered by the addition of genes
10	sticky end	J	synthetic chromosome consisting of tiny fragments of DNA integrated into a host cell
11	recombinant yeast cell	K	DNA which has been altered so that it has had genes added or removed

A student carried out an investigation into the production of an enzyme, normally produced in a eukaryotic cell, by a bacterium which had been transformed using recombinant DNA technology.

The bacterium was grown in a fermenter with glucose supplied as the food source.

The enzyme is released into the growth medium and is then harvested for human use.

Table 1 shows data for the change in enzyme concentration (units/cm³) over a period of 60 hours.

Graph 1 shows changes in the glucose concentration (g/100 cm³) and the dry mass of the bacterial cells (g/L) over the same period.

Table 1

Time (hours)	0	10	20	30	40	50	60
Enzyme concentration (units/cm³)	0	2	8	20	32	48	52

Graph 1

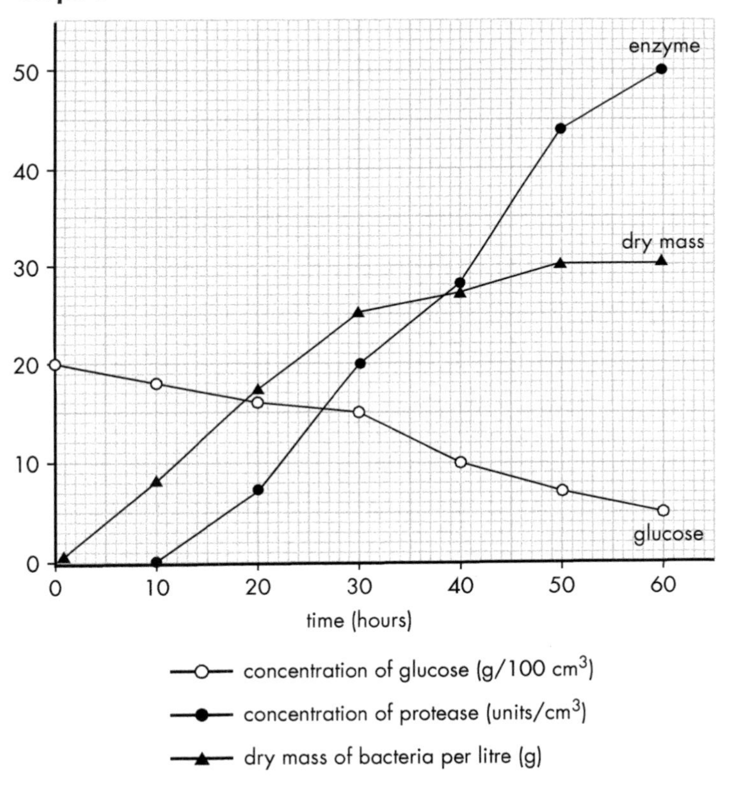

a Calculate the average decrease in glucose concentration per hour over the 60 hour period.

b State which 10-hour period gave rise to the greatest production of the enzyme.

c Calculate what percentage of glucose had been used up after 40 hours.

d State the whole number ratio of the bacterial dry masses at 10, 30 and 50 hours.

e State the independent variable in this investigation.

Hint: The independent variable is plotted on the *x*-axis.

16 Food supply and photosynthesis

16A Food supply

1. Explain what is meant by the term food security.

2. Copy and complete the following paragraph, which relates to global demand for food, by inserting one suitable term in each space provided.

   A word may be used once, more than once or not at all.

high – fall – community – population – increasing – decreasing – rise – low increases – maize – sugar – staple – export – supply – import – decreases

   The demand for food worldwide is __________ as the human __________ continues to _____. Current demand for food is greater than the ________. People who live in strong economic environments produce or can ___________ adequate supplies of food to supply their needs at all times. As other economies grow, their consumption of _______ foods such as rice, _________ and barley _____________ as their demand for higher value food ____________. Production of meat puts further strain on food supply as well as having a _______ energy demand.

3. Which of the following would both improve sustainable food production?

   A Reduced water logging and increased soil erosion

   B Optimum levels of fertiliser applied and reduction in the use of fossil fuels

   C Using chemical pesticides and high yielding genetic cultivars

   D Growing pest-resistant cultivars and increasing use of single crop fields

4. Decide if each of the following statements relating to sustainable food production in the table below is **TRUE** or **FALSE**. If the statement is **FALSE**, write the correct term to replace the term underlined in the statement.

Statement	True	False	Correction
Producing food in a sustainable way aims to decrease crop yields while conserving the natural resources on which the agriculture depends.			
Using fewer fertilisers and pesticides will reduce biodiversity.			
Sustainable food production maintains the mineral content of the soil.			
Sustainable food production systems are less labour intensive than non-sustainable food production systems.			

5. State the term used for a genetic variety of plants which has been selected for a desirable and stable phenotype.

6. **a** Explain why the application of a nitrogen-rich fertiliser can increase the growth of crop plants.

   b State one disadvantage of using chemical fertilisers.

7. Copy the following diagram. Connect the two different types of competition to the correct descriptions below by drawing a line between them.

	growing wheat plants need space, water and nutrients
interspecific competition	reduced by spacing crop seeds at time of planting
intraspecific competition	weeds and oats growing together
	reduced by application of a selective herbicide

8. Explain why livestock production is less energy efficient than crop production.

9. Goats have strong tongues and jaw muscles and can graze on very short grass and a wide variety of vegetation.

   Suggest an advantage to a farmer to raise goats as livestock.

16B Photosynthesis

1. The diagram below shows what happens when sunlight strikes a leaf

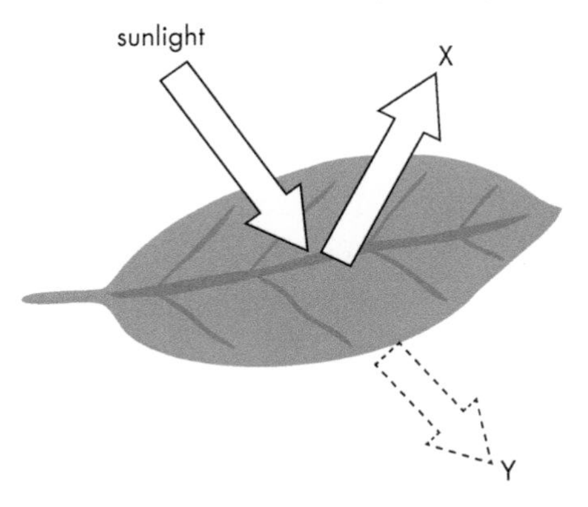

Which of the following correctly describes what happens to the light energy represented by X and Y?

	X	Y
A	transmission	absorption
B	reflection	transmission
C	absorption	reflection
D	transmission	reflection

2 The diagram below shows the different wavelengths of light striking a leaf. Explain why the leaf appears green to the eye.

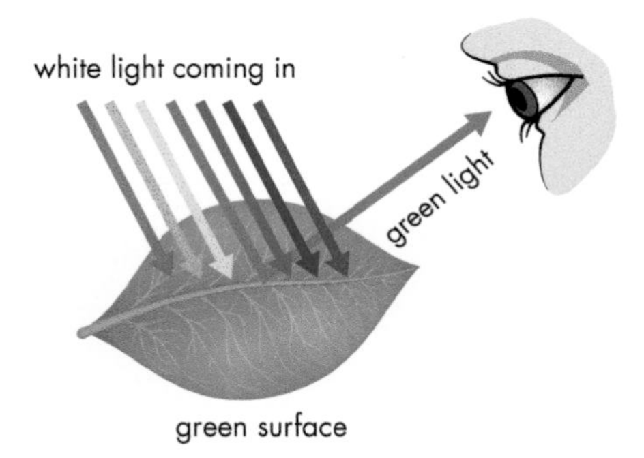

3 Decide if each of the following statements relating to photosynthetic pigments in the table below is **TRUE** or **FALSE**. If the statement is **FALSE**, write the correct term to replace the term underlined in the statement.

Statement	True	False	Correction
Photosynthetic pigments are located in the chloroplast.			
Carotenoids restrict the wavelengths of light which can be absorbed for photosynthesis.			

4 The graph below shows the absorption spectrum of a photosynthetic pigment and the rate of photosynthesis by a green plant over a range of colours of light.

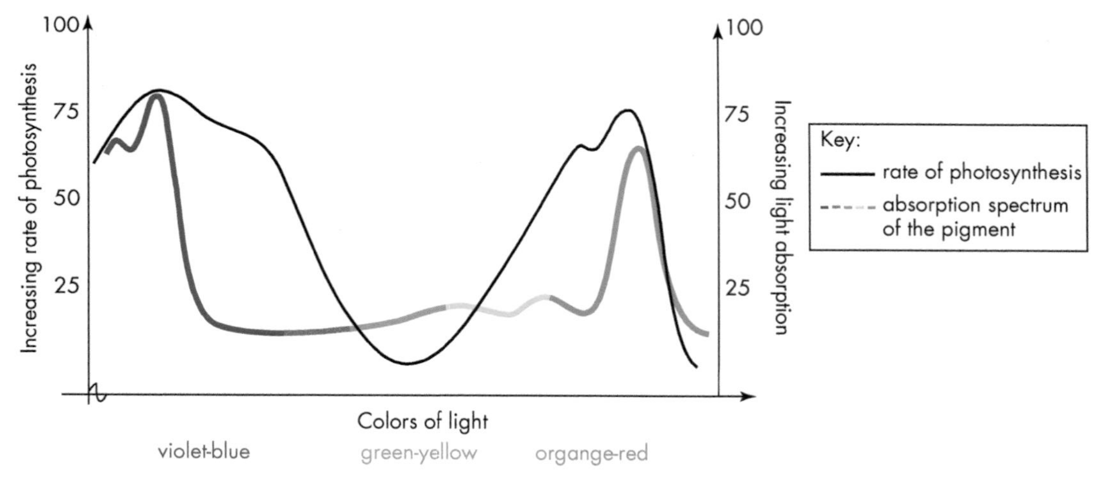

a Name photosynthetic pigment which would show this absorption spectrum.

b Explain how the graph shows that other pigments are involved in photosynthesis.

c Explain the advantage of plants having several different pigments in their chloroplasts.

5 Name of a group of accessory pigments and give their function.

6 The action spectrum for photosynthesis shows how green plants:

A use a wide range of wavelengths to photosynthesise

B split water using light energy in the process of photolysis

C absorb a wide range of wavelengths to photosynthesise

D absorb a limited range of wavelengths to photosynthesise.

7 Copy and complete the following paragraph, which relates to the biochemistry of photosynthesis, by inserting one suitable term in each space provided.

A word may be used once, more than once or not at all.

hydrogen – absorbed – ribosomes – chloroplasts – reflected – split – oxygen synthase – photolysis – pigments – NADH – stomata – coenzyme NADP – electrons – electron transport chain

In the _____________, light energy is __________ and used to _____ water to release oxygen and __________, a process called __________. Oxygen diffuses into the air through the __________ while the __________ is picked up by a __________ called ________ which is converted to ________. Absorbed light excites __________ in the ___________ within the ___________ which are passed along an ________ _________ ___________ to generate ATP using the enzyme ATP ________.

8 Decide if each of the following statements relating to the light stage of photosynthesis in the table below is **TRUE** or **FALSE**. If the statement is **FALSE**, write the correct term to replace the term underlined in the statement.

Statement	True	False	Correction
The splitting of water using light energy is called photolysis.			
The hydrogen produced by photolysis is picked up by a coenzyme.			
Electrons are passed along the electron transport chain releasing energy to produce ADP.			

9 The diagram below shows some parts of the light stage of photosynthesis.

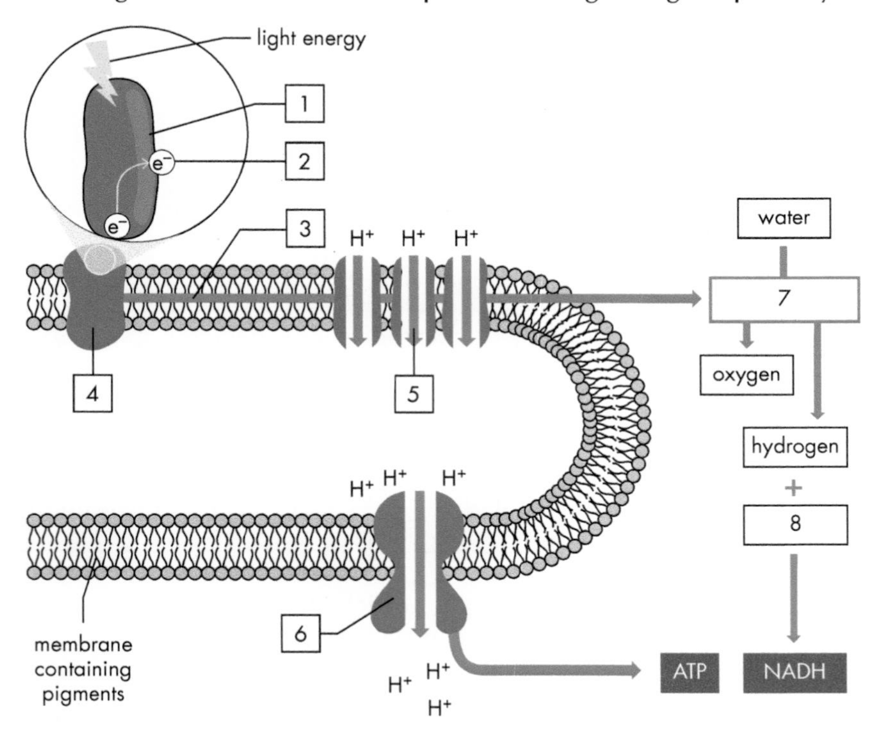

Identify which numbered boxes refer to each of the following:

a photolysis

b electrons picked up by first electron acceptor

c electrons passed along transport chain

d electrons become excited

e NADP

f hydrogen used to drive ATP synthase

g photosynthetic pigments absorb light energy

h hydrogen ions pumped across membrane.

10 The following statements refer to photosynthesis.

1 RuBisCO attaches carbon dioxide to RuBP.

2 3-phosphoglycerate is a precursor of glyceraldehyde-3-phosphate.

3 Hydrogen is picked up by NADP.

Which of the statements refers to the Calvin cycle?

A 1 and 2 only

B 2 and 3 only

C 1 and 3 only

D 1, 2 and 3

11 A number of chloroplasts were extracted from the cells of a nettle plant.

The chloroplasts were suspended in a solution of a dye which changes from blue to clear when it picks up hydrogen.

The chloroplasts and the dye were exposed to bright light for 15 minutes and the results are shown below.

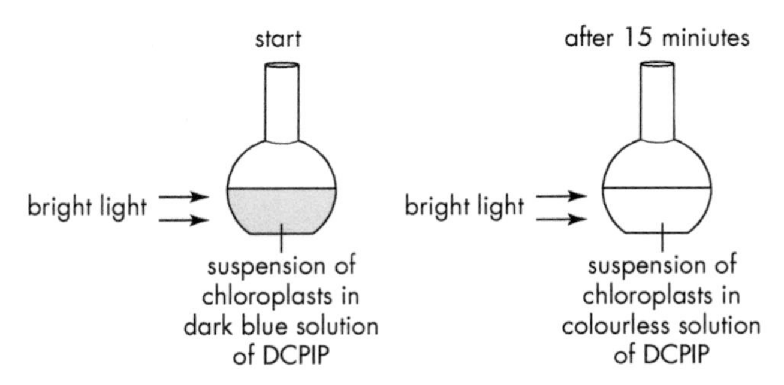

Explain what causes this colour change to take place.

12 Identify the compound or description in the table below which is linked to some of the metabolites of the Calvin cycle.

Compound	Description
	precursor of glyceraldehyde-3-phosphate
RuBisCO	
	first stable compound formed after carbon dioxide enters the cycle
G3P	

13 Some G3P may be used to form a carbohydrate.

a Name this carbohydrate.

b Give three possible uses of the glucose produced.

14 Which of the following would result when a plant which is photosynthesising freely is moved from bright light into a dark environment?

	Concentration of RuBP	Concentration of G3P
A	decreases	increases
B	increases	decreases
C	decreases	decreases
D	increases	increases

16C Short extended response question

Give an account of food security.

16D Longer extended response question

Write notes on the Calvin cycle.

> **Hint** This is typical of a question usually worth seven or more marks in the Higher Examination, so aim to give at least the same number of points, which must be relevant to the question. You can also make use of a suitably labelled diagram for this question.

16E Key terms

Link each key term below with the correct description.

	Key term		Description
1	food supply	**A**	location and affordability of food
2	sustainability	**B**	different wavelengths of light absorbed by pigments
3	soil erosion	**C**	shows rate of photosynthesis at each wavelength
4	cultivar	**D**	substances which help an enzyme to work
5	intraspecific competition	**E**	enzyme which attaches carbon dioxide to RuBP
6	interspecific competition	**F**	intermediate compound formed from 3-phosphoglycerate
7	photolysis	**G**	accessory pigments
8	carotenoids	**H**	competition between different species
9	absorption spectrum	**I**	competition between the same species
10	action spectrum	**J**	genetic plant variety which is selected and maintained in cultivation
11	coenzyme	**K**	loss of soil particles due to wind or rain
12	RuBisCo	**L**	not depleting natural resources
13	food security	**M**	availability of food
14	G3P	**N**	splitting of water using light energy

16F Data handling and experimental design

An investigation was carried out to study the effect of light intensity (units) on the rate of photosynthesis (units) using discs cut from a geranium plant.

The results are shown in the table below.

Light intensity (units)	10	20	30	40	50	60
Rate of photosynthesis (units)	5	25	60	90	95	95

a Plot a line graph to show these data.

b Using the data, predict the rate of photosynthesis at a light intensity of 65 units.

c State the independent variable in this investigation.

2 The graph below shows the effect of increasing light intensity (units) on the rate of photosynthesis (units) at different temperatures.

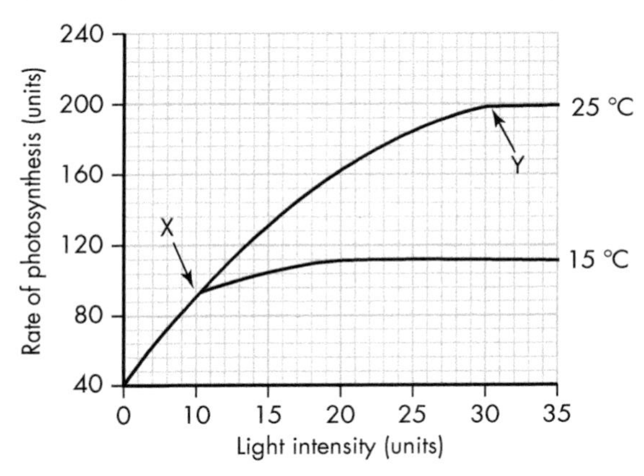

Which of the following is a valid conclusion from these data?

a At point X, temperature is a limiting factor preventing an increase in the rate of photosynthesis.

b An increase in temperature has little effect at very low light intensities.

c Only at light intensities greater than 20 units does temperature affect the rate of photosynthesis.

d At point Y, light intensity is limiting the rate of photosynthesis.

3 An ecosystem receives 2 000 000 kJ/m²/year of sunlight energy. 95% of this energy is not used in photosynthesis.

Calculate the energy trapped by the producers in this ecosystem.

17 Plant and animal breeding

17A Plant and animal breeding

1. Using the word bank provided, copy and complete the following description of the selection of genetic characteristics in animals and plants.

   A word may be used once, more than once or not at all.

breeding – cultivar – productivity – traits – phenotypes – genotypes – genes – specific

   Farmers select genetic ________ which give rise to desirable ________ such as increased ____________ of animals and plants. This ensures the ________ are conserved in future generations. Only those individuals whose ___________ give rise to the desired ___________ are selected to be included in __________ programmes. By breeding only those organisms which possess a _______ desirable trait, an improved __________ can be sustained.

2. Copy and complete the table below suggesting three desirable genetic characteristics in livestock and crops.

Livestock	Crops

3. Which of the following statements is true?

   a Resistance of wheat to fungal infection is not a desirable phenotype.

   b A plant cultivar can be obtained by selection of desirable phenotypes.

   c Productivity cannot be improved by selective breeding.

   d Selective breeding requires few generations to produce desirable phenotypes.

4. Explain why inbreeding may cause harmful recessive traits to emerge.

5. State one advantage of selective breeding.

6. The following are four stages involved in selective breeding.

   1 Offspring produced which have the desired phenotype

   2 Offspring with desired phenotype are bred

   3 Repeat over many generations

   4 Close individuals with desired phenotypes are bred together

   The correct order of these stages is:

   A 1→2→3→4

   B 2→3→4→1

   C 3→4→2→1

   D 4→1→2→3

17B Plant field trials

1 Decide if each of the following statements relating to construction of a field trial in the table below is **TRUE** or **FALSE**. If the statement is **FALSE**, write the correct term to replace the term underlined in the statement.

Statement	True	False	Correction
Multiple replicates are needed due to potential variation within the trial area.			
Field trials cannot generate data on the effect of applying different levels of fertiliser.			

2 A field trial to test the effect of a new selective herbicide is about to be carried out.

a Give one variable which must be controlled to make the results of the field trial valid.

b State one control measure which needs to be put in.

3 In designing a field trial, which of the following will not increase the validity of the results?

a Randomisation of treatments

b Selection of treatments

c Number of replicates

d Cost of the treatment

17C Inbreeding

1 Decide if each of the following statements relating to selecting and breeding in the table below is TRUE or FALSE. If the statement is FALSE, write the correct term to replace the term underlined in the statement.

Statement	True	False	Correction
Animals and cross-pollinating plants are naturally inbreeding.			
Animals and cross-pollinating plants which have been inbred show a narrow range of genetic variation.			
Heterozygous individuals generally are more vigorous and have a good rate of growth.			

2 Which of the following descriptions below apply to inbreeding?

1. mating genetically close relatives for several generations

2. reduces genetic variation in offspring

3. increases genetic variation in offspring

4. mating distantly related individuals for several generations

5. offspring likely to have reduced biological fitness

A 1 and 2

B 2 and 5

C 3 and 4

D 3 and 5

3 Explain why inbreeding depression is more likely to occur in small populations compared with large populations.

4 The diagram below shows the results of some research into the inheritance pattern of an adder. A recessive allele, a, causes stillbirths and deformities in the offspring.

A = dominant recessive allele a = recessive harmful allele

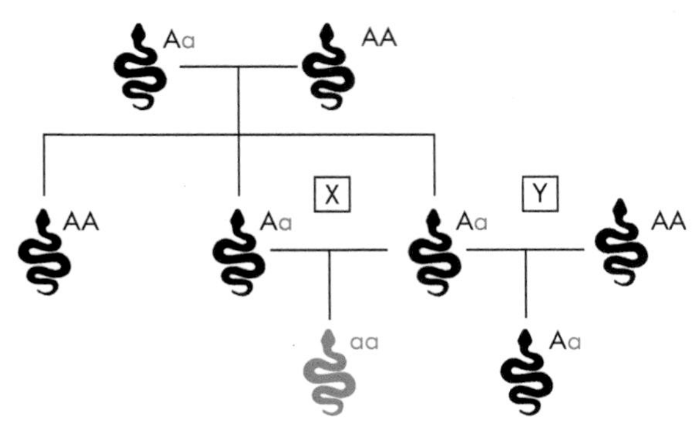

a Using the appropriate letter, state where inbreeding has taken place.

b Suggest one way researchers could help reduce inbreeding depression.

17D Cross breeding and F1 hybrids

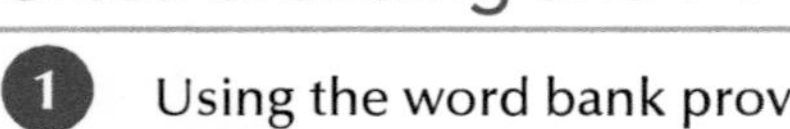

1 Using the word bank provided, copy and complete the following description of a monohybrid cross.

A word may be used once, more than once or not at all.

multiple – monohybrid – alleles – parents – gene – cross – single – parent – gamete characteristics – breeding – F_2 – dominant – offspring – generation – F_1 – population

A ______ between two __________ that differ in their _______ for one specific _____ is called a _________ or _______-factor _____. For example, one ________ might be homozygous ____________ and the other homozygous recessive. The ________ from this cross, called the F_1 ___________, will possess one dominant and one recessive form of the ______. If two F_1 are crossed together, the offspring are called the F_2 generation. Farmers make use of this when ______ ________________ different varieties of an animal to produce a new _________ with improved _____________.

2. Explain why plants used as food crops may be inbred.

3. Explain why the F_2 generation from a cross between two F_1 hybrids may be of little use for further production.

17E Genetic technology

1. Using the word bank provided, copy and complete the following description of the application of genetic technology to improve the phenotype of organisms.

   A word may be used once, more than once or not at all.

identified – polymerase chain reaction – desirable – genome sequencing transformation – breeding – genome – breeding programmes

Using gene technology, such as ______________ ___________ _________ and ___________ _______________, organisms with ___________ genes can be ______________ and then used in ________________ programmes. Using genetic __________________ techniques a single gene can be inserted into a __________ which can then be used in ___________.

17F Short extended response question

Write notes on inbreeding depression.

17G Longer extended response question

Give an account of how productivity can be improved by genetic manipulation.

17H Key terms

Link each key term below with the correct description.

1. cultivar
2. genetically modified organism
3. hybrid vigour
4. inbreeding
5. inbreeding depression
6. allele
7. monohybrid cross
8. genomic sequencing

A sexual reproduction between closely related organisms

B cultivated variety of plant

C enhanced phenotype in a heterozygous individual

D animal or plant whose genetic material has been manipulated

E negative effect of continued inbreeding

F genetic cross involving one gene

G using PCR and bioinformatics to analyse all the genes present in an organism

H alternative form of a gene

17I Data handling and experimental design

The table below shows the results of an investigation into five different cultivars of a plant called *Alfafa* over a period of five years. 100 plants of each cultivar were used.

Cultivar	Seed germination (%)			Mass of 1000 seeds (g)			Seed moisture content (%)		
	Year when seeds were produced								
	2000	2005	Average	2000	2005	Average	2000	2005	Average
1	60	70	65	1.5		1.25	8.0	7.6	
2		70	63	1.3	1.5	1.40	7.5	7.9	7.7
3	10	66	38	1.6	1.4	1.50	6.8	7.0	
4	74	80	77	2.0	2.2	2.10	7.2	7.0	7.1
5	45		50	2.2		2.10	8.0	7.6	7.8

a Complete the table by calculating the missing data.

b State the independent variable in this investigation.

c Draw a bar chart to show the percentage seed germination and mass of 1000 seeds for cultivar 3 in 2000 and 2005.

Hint: When drawing bar charts, make sure you use more than 50% of the grid, make each bar the same width, space out bars equally, label each axis and include units where appropriate, and use both a ruler and a sharp pencil.

d State one feature of this investigation which helped make the results reliable.

e Calculate the ratio of the average seed germination for cultivars 1 and 5.

Hint: Practise calculating ratios and be able to express them as the lowest whole numbers. Watch the order of your answer matches that of the question.

f Identify the cultivar which had an average of 20% more mass per 1000 seeds compared with cultivar 1.

2 The table below shows the average yield (tonnes/hectare) for two years, 1880 and 1980, for four different crop plants.

Crop	Average yield (tonnes/hectare)	
	1880	1980
Oats	1.9	3.8
Potato	13	39
Barley	2.0	6.6
Turnip	3.2	6.8

a Identify which crop showed the smallest percentage increase in average yield/hectare over the two years.

b In 1980 a farmer grew 20 hectares of turnip but lost 10% of the crop due to disease. Calculate his final yield in tonnes.

18 Crop protection

18A Weeds

1. Suggest one reason why modern crop agriculture often favours the growth of weeds.
2. Suggest one beneficial effect of weed growth.
3. State three resources which weeds compete with crops for.
4. Decide if each of the following statements relating to properties of weeds in the table below is **TRUE** or **FALSE**. If the statement is **FALSE**, write the correct term to replace the term underlined in the statement.

Statement	True	False	Correction
The seeds of annual weeds germinate slowly in response to light or recent soil disturbance.			
Perennial weeds produce very high numbers of seeds.			
The seeds of weeds often have mechanisms for dormancy which allows them to remain viable for a long time.			
Many weeds are unable to grow by vegetative reproduction.			

5. Dandelion is a common weed which has a very strong taproot system which can withstand severe environmental stress.

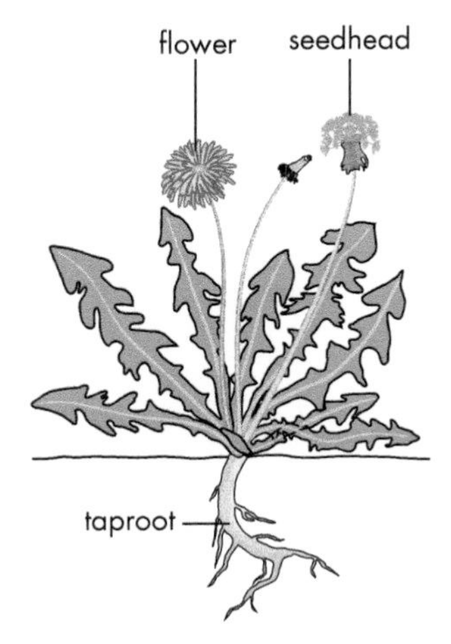

a Suggest if this is likely to be a perennial or annual weed and suggest a reason for your answer.

b State what type of herbicide could be needed to eliminate dandelion from a garden and give a reason for your answer.

c Give three disadvantages of using a chemical weedkiller to control dandelions.

d The seedhead produces masses of parachute-like seeds.

Suggest how this helps spread the dandelion.

6. Which of the following two descriptions apply to perennial weeds?

A Reproduce sexually and have long life cycles

B Have short life cycles and produce large numbers of seeds

C Produce storage organs and reproduce vegetatively

D Grow rapidly and reproduce sexually

7. Decide if each of the following statements relating to properties of crop pests in the table below is **TRUE** or **FALSE**. If the statement is **FALSE**, write the correct term to replace the term underlined in the statement.

Statement	True	False	Correction
Most crop pests are small vertebrates.			
Plant diseases can be caused by fungi, bacteria or viruses, which are often carried by invertebrates.			

8. Give three examples of common crop pests.

9. Which of the following are NOT common potential causes of disease in crops?

A Protozoa

B Viruses

C Fungi

D Bacteria

18B Control of weeds, pests and diseases by cultural means

a State one way in which cultivation techniques can be used to prevent the build up of weeds, pests and diseases.

b Explain how this technique is effective.

18C Control of weeds, pests and diseases by chemical means

1. Copy the following diagram. Two different herbicides are shown. Link them to the correct description by drawing a line between them.

selective herbicide	absorbed into phloem and xylem
systemic herbicide	attacks only broad-leaved plants

2 Using the word bank provided, copy and complete the following description of pesticides and fungicides.

A word may be used once, more than once or not at all.

more – herbicides – target – toxic – species – insects pests – bacteria – specific – promote – bioaccumulate degrade – persistent – fungicides – inhibit – systemic – less

A pesticide is a chemical which controls ______ in and on plants. Such pests include ___________, fungi and viruses as well as molluscs, _________ and certain types of worms. Although pesticides have many benefits, some have drawbacks as well such as potential ________ effects on humans and other _________. Most pesticides are, however, very ________ to their ________ pest. Some pesticides are ____________ and do not ________ easily or quickly. Others can ___________ or be magnified in food chains. ___________ are chemicals which are applied to growing crops to kill or _________ the growth of pathogenic fungi. Like pesticides, ___________ can be contact, _________ or selective. Protection of a crop, by early spraying, is much _____ effective than treatment after infection.

3 Which of the following is true?

A Systemic fungicides do not penetrate xylem and phloem.

B Insect pests can develop resistance to pesticides through mutation.

C Herbicides kill or inhibit invertebrate pests.

D Selective pesticides kill a range of animal pests.

18D Biological control and integrated pest management

1 A farmer has a problem with a particular crop pest which has a natural insect predator. She decides to alter the environment within which the crop is grown by arranging different plantings near and around the crops as shown below to attract the predator.

a State the type of pest control this farmer is using.

b Suggest why she has arranged the different plantings near and around the crops.

c State two benefits and two limitations of this type of pest control which is not based on the use of chemicals.

2 Biological control is:

A using living organisms in sewage breakdown

B regulating bird populations by designating protected areas

C using a natural predator or parasite of the pest

D reducing soil loss by planting native trees.

3 Decide if each of the following statements relating to biological control in the table below is **TRUE** or **FALSE**. If the statement is **FALSE**, write the correct term to replace the term underlined in the statement.

Statement	True	False	Correction
The negative impact of chemical control is greatly reduced using biological control.			
The pest species is the secondary food source for the predator species.			
Natural predators used in biological control may themselves become pests.			

18E Short extended response question

Discuss cultural methods of controlling weed growth.

18F Longer extended response question

Give an account of different ways of protecting crops.

18G Key terms

Link each key term below with the correct description.

1 competition

2 weed

3 perennial weed

4 herbicide

5 annual weed

6 crop rotation

7 systemic pesticide

8 fungicide

9 selective herbicide

10 bioaccumulation

11 integrated pest management

A undesirable plant

B organisms struggle to obtain similar resources

C weed with short life cycle

D cultural strategy for controlling weeds

E weed with long life cycle

F chemical which kills plant pests

G chemical which kills pathogenic fungi

H combined strategy for controlling pests

I chemical which kills and is absorbed into body of pest

J chemical which kills some, but not all, plant pests

K increasing concentration of a chemical along a food chain

18H Data handling and experimental design

1 The percentage increases in pesticide used in seven different countries from 1990 until 2010 are shown in the table below.

Country	% increase
1	820
2	500
3	350
4	300
5	260
6	250
7	150

a Draw a bar graph of these data.

b State the independent variable.

c Calculate the ratio of the percentages for countries 4, 3 and 2.

2 The bar chart below shows the percentage loss in yield of seven different crops per year across the world in 1990.

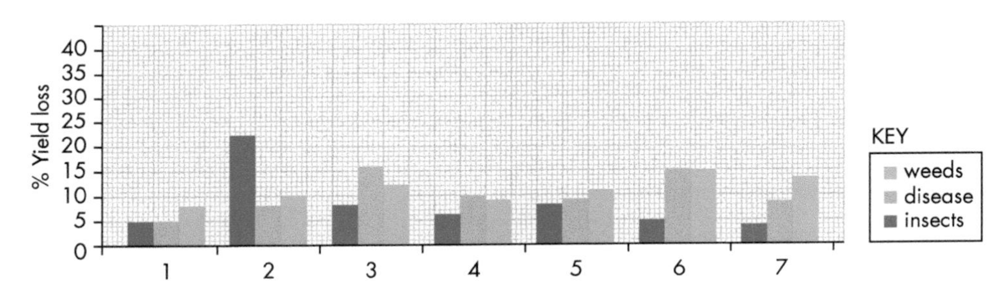

a Identify the crop(s) that had the highest annual loss in yield due to:

i disease

ii weeds.

b Identify the crop that had the lowest annual loss in yield due to insects.

c Calculate the whole number ratio of the percentage annual loss in yield in rice due to disease, insects and weeds.

d Identify which crop would give the greatest increase in the percentage annual yield when treated with:

i insecticide

ii selective weedkiller.

19 Animal welfare

19A Livestock production

1. Explain how good animal welfare can increase productivity.

2. Decide if each of the following statements relating to animal welfare conditions in the table below is **TRUE** or **FALSE**. If the statement is **FALSE**, write the correct term to replace the term underlined in the statement.

Statement	True	False	Correction
Stereotypic behaviour is seen as repeated patterns of purposeless behaviour.			
Misdirected behaviour is seen as normal behaviour directed against an animal itself.			
An ethogram cannot confirm to a farmer that the environmental conditions are meeting the needs of his animals.			

3. Which of the following is least likely to develop altered behaviours in animals?
   - **a** Elephants housed in confined zoo enclosure
   - **b** Sheep allowed to hill graze all year
   - **c** Pigs reared in sow cages
   - **d** Pet birds housed in small cages

4. Using the word bank provided, copy and complete the following paragraph, which relates to misdirected behaviour.

   A word may be used once, more than once or not at all.

beaks – environment – abnormal – misdirected – normal – feathers hyper-aggression – reduced – increased – tails

   __________ behaviour is when __________ behaviour is expressed in a different or inappropriate situation towards itself or others. This includes: birds over-plucking their _________, pigs biting their _____ or biting on solid objects, ______________, and _________ reproductive success.

19B Observing animal behaviour

The following is a partially completed ethogram for a dog. Using the letters of the phrases provided, select those best suited to complete the table.

Not all the letters will be used.

a shows fear when offered food

b unwanted behaviours

c defecates normally

d reacts badly to walking

e walks normally

Sociable behaviours	1
feeds properly and can be handled	protects food and growls when approached during feeding
2	eats faeces
3	**4**

Explain how a preference test can be used to help improve animal welfare conditions.

19C Short extended response question

Give an account of stereotypy.

19D Longer extended response question

Discuss motivation in animals.

19E Key terms

Link each key term below with the correct description.

1	animal welfare	**A**	farming to produce as much output as possible
2	intensive farming	**B**	study of behaviour
3	altered levels of activity	**C**	experiment involving choices of conditions
4	stereotypy	**D**	checklist of animal behaviours
5	misdirected behaviour	**E**	abnormal level of frustration
6	ethology	**F**	repetitive behaviour with no obvious motivation
7	ethogram	**G**	normal behaviour displayed in an unusual situation
8	preference test	**H**	drive to satisfy a need
9	motivation	**I**	well-being of animals
10	failure in sexual behaviour	**J**	lack of normal care of offspring
11	failure in parental behaviour	**K**	inability to carry out normal sexual activity

19F Data handling and experimental design

1 The pie chart below shows the percentage time spent by a sea lion performing different behaviours.

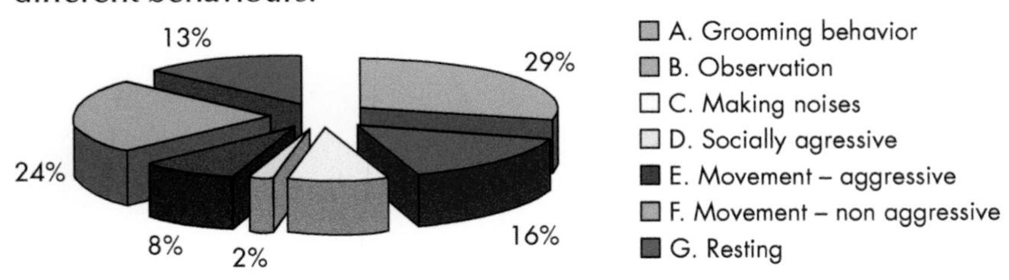

Hint Make sure you are familiar with the different types of data presentation, which include line graphs, bar charts and pie charts.

a Calculate the percentage time the sea lion spends making noises.

b Identify the most common behaviour.

c Identify which behaviour is twice as frequent as aggressive movement.

2 The following bar chart shows data collected by an observer of a monkey in a zoo.

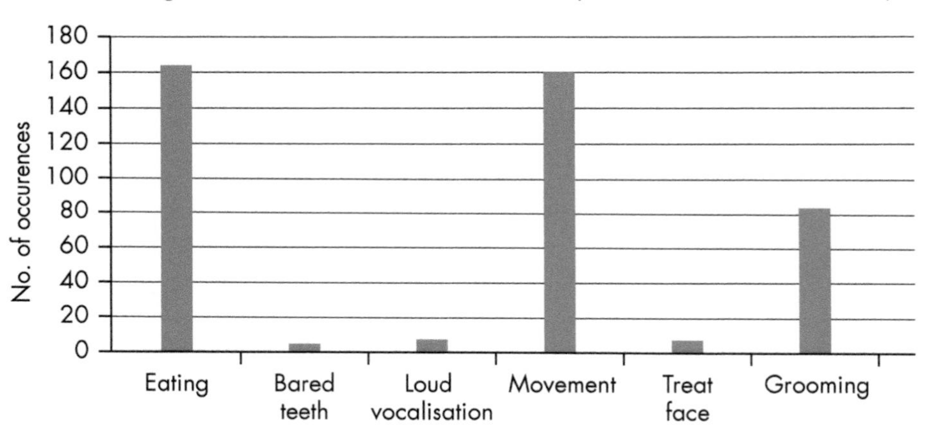

a Identify the most common behaviour and state the number of occurrences.

b Calculate how many more occurrences there were of grooming compared with bared teeth.

c State one precaution the observer should take to improve the validity of the results.

d State an aspect of the investigation which may not have ensured that reliable results were obtained and explain how it could be addressed.

20 Symbiosis

20A Symbiosis

1
a State what is meant by the term symbiosis.
b State two types of symbiosis.

20B Parasitic relationships

1
a State what is meant by parasitism.
b Give an example of a parasite.

2
State the name of the transmission method:
a involving physical contact
b where the parasite is passed on by another organism
c involving survival in adverse conditions.

3
Explain why parasites often cannot survive without their host.

20C Mutualism

1
a State what is meant by mutualism.
b Describe an example of mutualism.

2
a Explain the process by which mutualism facilitated the evolution of mitochondria and chloroplasts into larger cells.
b Describe the main mutualistic benefits to this process.

3
Decide if each of the following statements relating to symbiosis in the table below is **TRUE** or **FALSE**. If the statement is **FALSE**, write the correct term to replace the term underlined in the statement.

Statement	True	False	Correction
Symbiosis is described as co-evolved intimate relationships between members of two different species.			
A parasite loses in terms of energy or nutrients, whereas its host is harmed by the loss of these resources.			
Both mutualistic partner species benefit in an interdependent relationship.			

20D Longer extended response question

Give an account of parasitism and mutualism as types of symbiosis.

20E Key terms

Link each key term below with the correct description.

	Key term		Description
1	symbiosis	A	organism which relies upon a host to survive
2	parasitism	B	organism which plays an indirect role in the parasitic life cycle
3	parasite	C	organism that transmits parasites from host to host
4	transmission	D	relationship where both species benefit
5	vector	E	relationship where one organism benefits in terms of energy, whereas the other loses
6	host	F	co-evolved intimate relationships between members of two different species
7	direct contact	G	method of transmission involving physical contact between two hosts
8	resistant stage	H	term for transferring parasites from host to host
9	secondary host	I	organism where adult parasites reside
10	mutualism	J	stage of parasitic life cycle where eggs/larvae can survive adverse conditions until entering a new host

20F Data handling and experimental design

A scientist carried out an investigation on how temperature affects the rate of parasite egg hatching during their resistant stage.

She exposed 300 parasite eggs to three different temperatures (100 eggs per temperature used) and observed how many hatched after 10 days exposure to that temperature.

The results are shown in the table below.

Day	Number of eggs unhatched		
	5 °C	37 °C	60 °C
0	100	100	100
1	92	81	98
2	83	54	95
3	71	32	94
4	59	13	94
5	52	7	94
6	48	5	94
7	41	4	94
8	35	2	93
9	31	2	92
10	28	2	90

a **i** Name the independent variable in this experiment.

ii Identify a variable, not already mentioned, that should be kept constant so that a valid conclusion can be drawn.

b Suggest how the temperature for each experiment could be kept constant.

c State an aspect of the investigation which helped to ensure that more reliable results were obtained.

d Plot a line graph to show the results of the investigation.

e Draw a conclusion from these results.

21 Social behaviour

21A Social hierarchy

1 **a** State what is meant by the term social hierarchy.
b State the terms given to the highest and lowest ranked individuals in the hierarchy.
c Describe one example of social hierarchy.

Hint: Examples are important in this topic in order to fully contextualise many of the terms used.

21B Altruism and kin selection and its influence on survival

1 **a** State what is meant by the term altruism.
b Describe when altruism is most commonly seen in animals.

2 State what is meant by the term reciprocal altruism and give one example.

3 **a** State what is meant by the term kin selection.
b Describe the advantage of kin selection.

21C Social insects, the structure of their society and their importance

1 State two different roles found in social insects.

2 Name two species that are social insects.

3 Describe the function of the queens and drones in insect societies.

4 Describe the role carried out by the sterile workers.

21D Primate behaviour

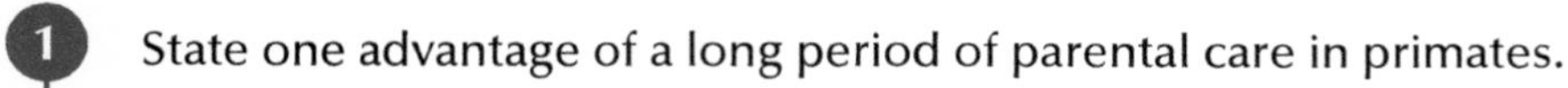

1 State one advantage of a long period of parental care in primates.

2 **a** State two complex behaviours that reduce unnecessary conflict.
b State two examples of these complex behaviours.

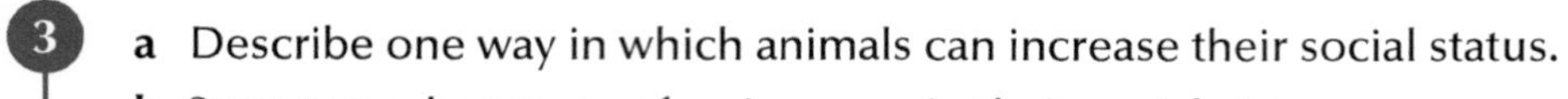

3 **a** Describe one way in which animals can increase their social status.
b State two advantages of an increase in their social status.

4 Decide if each of the following statements relating to social behaviour in the table below is **TRUE** or **FALSE**. If the statement is **FALSE**, write the correct term to replace the term underlined in the statement.

Statement	True	False	Correction
An altruistic behaviour harms the donor individual but benefits the recipient.			
Most members of a social insect colony are fertile workers who cooperate with close relatives to raise relatives.			

5 Copy and complete the following sentences by inserting one suitable term in each space provided.

A word may be used once, more than once or not at all.

grooming – complex – conflict – social – appeasement – parental – alliances

The long period of __________ care in primates gives an opportunity to learn __________social behaviours. To reduce unnecessary __________, social primates use ritualistic display and ____________ behaviours. __________, facial expression, body posture and sexual presentation are important in different species. In some monkeys and apes, ____________ form between individuals which are often used to increase _________ status within the group.

21E Short extended response question

Write notes on primate behaviour.

21F Longer extended response question

Give an account of co-operative hunting and social defence in animals.

21G Key terms

Link each key term below with the correct description.

	Key term		Description
1	social hierarchy	A	co-operation between members of the social group to hunt their prey
2	altruism	B	lower ranked animal in social hierarchy
3	reciprocal altruism	C	rank in social hierarchy
4	kin selection	D	animals that belong to an order of mammals including humans, apes and lemurs
5	social insects	E	staying together as a large group as a means of defence
6	primate	F	submissive display that is the reverse of a threat ritual
7	social status	G	threat ritual from one organism to another
8	social defence	H	the roles of donor and recipient later reverse
9	co-operative hunting	I	societies of e.g. ants, bees, wasps where only some organisms contribute reproductively
10	subordinate	J	altruism among animals that are closely related
11	ritualistic display	K	behaviour that harms the donor individual but benefits the recipient
12	appeasement behaviour	L	system where the members of a social group are organised into a graded order of rank

21H Data handling and experimental design

The graph below shows the net energy gain or loss from hunting and eating prey of different masses.

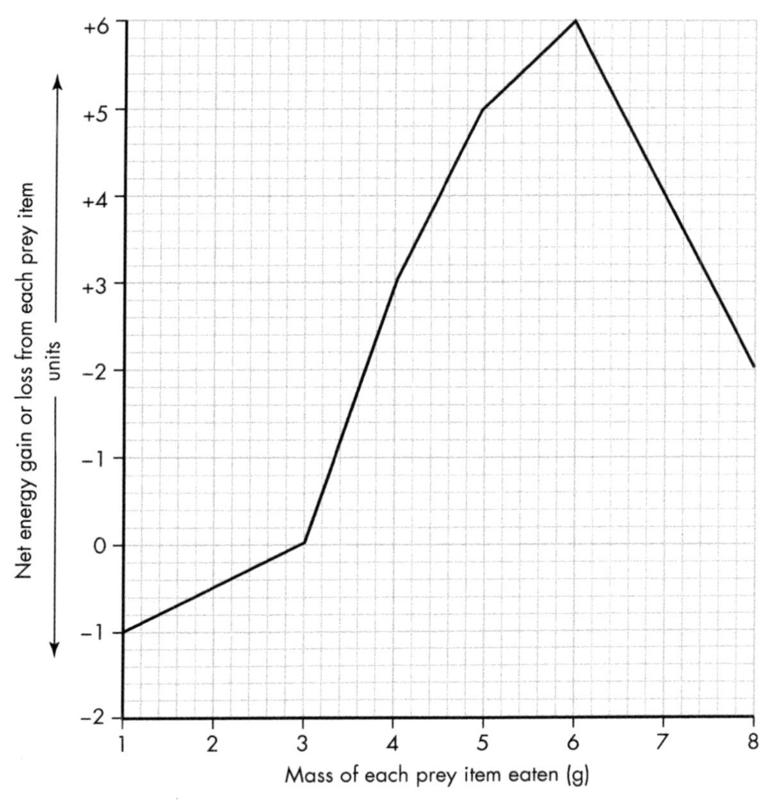

It can be concluded from the graph that:

A prey weighing between 1 g and 3 g are rarer than prey weighing between 3 g and 6 g

B hunting and eating prey weighing above 6 g involves a net energy loss

C prey weighing 8 g contain less energy than prey of mass 6 g

D hunting and eating prey weighing below 3 g involves a net energy loss.

2 A pack of wolves was studied hunting for three different prey species.

The table below shows the number of hunts carried out and the percentage of hunts that resulted in one kill.

Prey species	Number of hunts	Percentage of hunts resulting in one kill
Deer	50	10
Elk	100	25
Moose	100	20

Which of the following conclusions can be drawn from this data?

A Deer used social defence mechanisms most successfully.

B Co-operative hunting allows the wolves to hunt more successfully.

C Wolves killed more elk than any other prey species.

D More individual deer escaped than any other prey species.

3 A group of zoologists observed a group of primates over a 1-hour period, recording the number of times certain behaviours were displayed by each primate.

The results are shown in the table below.

Behaviour	Number of times behaviour displayed by each primate			
	A	B	C	D
Grooming	20	0	3	12
Sexual presentation	14	0	0	6
Chest beating	0	14	7	2
Showing teeth	1	23	12	4

a Calculate the number of times a ritualistic display was observed during the hour.

b Calculate the simplest whole number ratio between the total number of grooming behaviours observed and the total number of sexual presentation behaviour observed.

c Calculate the percentage increase in the number of times chest beating was observed between primates C and B.

d State the behaviour which was displayed the most times between all primates.

e Name the primate which is:

i the lowest ranked

ii the highest ranked

iii attempting to create the most alliances.

22 Components of biodiversity

22A Measuring biodiversity

1. State three measurable components of biodiversity.
2. Describe the impact upon biodiversity of one population becoming extinct.
3. State what is meant by the term genetic diversity.
4. State two components of species diversity.
5. Describe the impact of a dominant species upon a community.
6. State what is meant by the term ecosystem diversity.
7. Decide if each of the following statements relating to mass extinction and biodiversity in the table below is **TRUE** or **FALSE**. If the statement is **FALSE**, write the correct term to replace the term underlined in the statement.

Statement	True	False	Correction
Fossil evidence indicates that there have been several mass extinction events in the past.			
The extinction of mega fauna is correlated with the spread of humans.			
A community with a dominant species has a higher species diversity than one with the same species richness but no particularly dominant species.			

22B Longer extended response question

Give an account of how biodiversity is measured.

22C Key terms

Link each key term below with the correct description.

1. biodiversity
2. genetic diversity
3. species diversity
4. dominant species
5. ecosystem diversity

A measured by genetic diversity, species diversity and ecosystem diversity

B most numerous species in an ecosystem

C variation represented by the number and frequency of all the alleles in a population

D number of distinct ecosystems within a defined area

E the number of different species in an ecosystem and the proportion of each species in the ecosystem

22D Data handling and experimental design

The graph below shows the annual variation in the biomass and population density of the insect *Bourletiella viridescens*, found primarily in the Cairngorms in the UK.

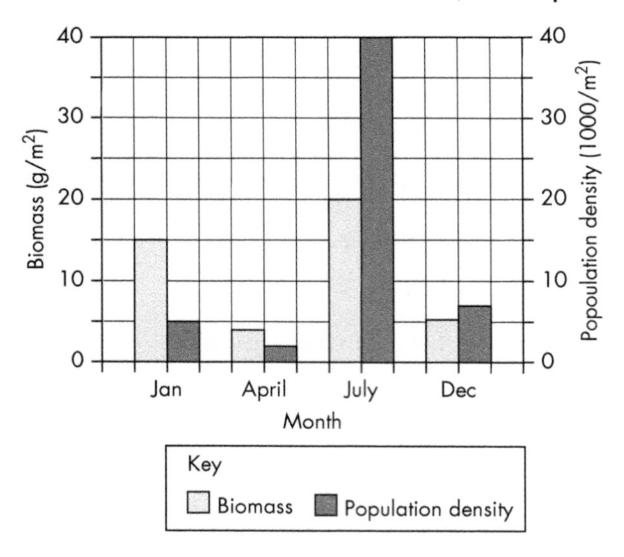

During which month do individual *Bourletiella viridescens* have the greatest average biomass?

A January

B April

C July

D December

The graph below shows changes in the number of breeding pairs of peregrine falcons between 1950 and 1990.

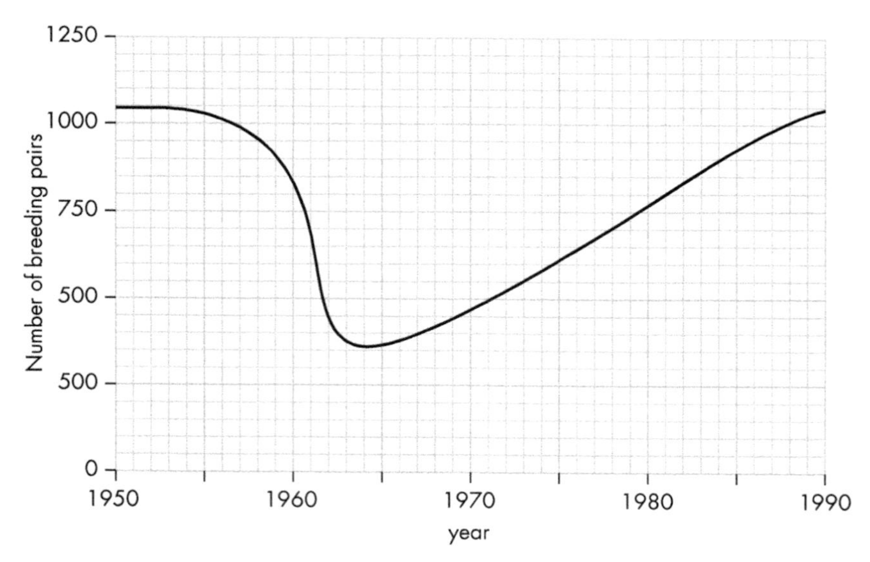

Which line in the table correctly shows the overall increase in the 10-year periods shown?

	10-year period	Increase in number of breeding pairs
A	1950–1960	250
B	1960–1970	350
C	1970–1980	300
D	1980–1990	400

3 A group of scientists carried out an investigation on the effect of habitat island area upon species richness.

They observed four different habitat islands and recorded the species richness for one 12-hour period. They also measured the area (m^2) of each habitat island.

The results recorded are in the table below.

	Habitat island			
	A	B	C	D
Area (m^2)	1200	6000	4000	10 000
Species richness	14	32	24	36

a **i** Name the independent variable in this experiment.

ii Identify two variables, not already mentioned, that should be kept constant so that a valid conclusion can be drawn.

b Describe the relationship between the area of the habitat island and the species richness.

c State how the procedure could be improved to increase the reliability of the results.

d Plot a line graph to show the results of the investigation.

e Draw a conclusion from these results.

f Suggest why it would not be valid to compare the species richness between the habitat islands.

g Suggest how you could compare the species richness between the islands in a valid way.

h Calculate the percentage increase in area between habitat island C and B.

i Calculate the simplest whole number ratio between the area of habitat island A and B.

j Predict the species richness of a habitat island with an area of 8000 m^2.

23 Threats to biodiversity

23A Exploitation and recovery of populations and the impact on their genetic diversity

1. State three examples of human exploitation which cause a drastic reduction in numbers of organisms.
2. Describe the impact on the genetic diversity of species when they have been exploited.
3. State the term used to describe when small populations lose the genetic variation necessary to enable evolutionary responses to environmental change.
4. Describe one way in which this loss of genetic diversity can critically impact a population.
5. Describe an example of naturally low genetic diversity and the impact on the species.

23B Habitat loss, habitat fragments and their impact on species richness

1. Describe two ways in which habitats can be lost due to human activities.
2. Describe what is meant by habitat fragmentation.
3. Describe the consequences of habitat fragmentation upon the habitat size and the species within it.
4. **a** Describe one method used to overcome isolated fragments of habitat.
   b Explain the main advantages of this method.

23C Introduced, naturalised and invasive species and their impact on indigenous populations

1. State what is meant by the terms:
   a introduced species
   b naturalised species
   c invasive species.

   > Hint: Students often confuse these three terms – learning an example of each can often help you distinguish between them in exams.

2. Name factors that may not limit invasive species in their new habitat.
3. Describe three impacts that invasive species may have upon native species.

4. Decide if each of the following statements relating to threats to biodiversity in the table below is **TRUE** or **FALSE**. If the statement is **FALSE**, write the correct term to replace the term underlined in the statement.

Statement	True	False	Correction
Inbreeding in large populations results in poor reproductive rates.			
Some species have a naturally high genetic diversity in their population and yet remain viable.			

5. Copy and complete the following sentences by inserting one suitable term in each space provided.

A word may be used once, more than once or not at all.

competitors – native – predators – hybridise – prey – rapidly – resources

Invasive species are naturalised species that spread __________ and eliminate __________ species. Invasive species may well be free of the __________, parasites, pathogens and __________ that limit their population in their native habitat. They may _____ on native species, out-compete them for __________ or __________ with them.

23D Short extended response question

Write notes on habitat loss and fragments.

23E Longer extended response question

Give an account of invasive, naturalised and introduced species.

23F Key terms

Link each key term below with the correct description.

1. exploitation
2. bottleneck effect
3. habitat loss
4. species richness
5. human activities
6. habitat corridor
7. introduced species
8. naturalised species
9. invasive species
10. native species
11. edge species

A species adapted to the habitat edges

B species that humans have moved either intentionally or accidentally to new geographic locations

C link isolated fragments of habitat

D naturalised species that spread rapidly and eliminate native species

E species that naturally habitats an area without human intervention

H action carried out by humans which negatively impact an ecosystem/organism

I humans benefitting from resources at other organisms' expense

J number of different species in an ecosystem

K reduction in habitat area

L organisms that become established within wild communities

M small populations lose the genetic variation necessary to enable evolutionary responses to environmental change

23G Data handling and experimental design

1. The graph below shows the changes in the populations of red and grey squirrels in an area of woodland over a 10-year period.

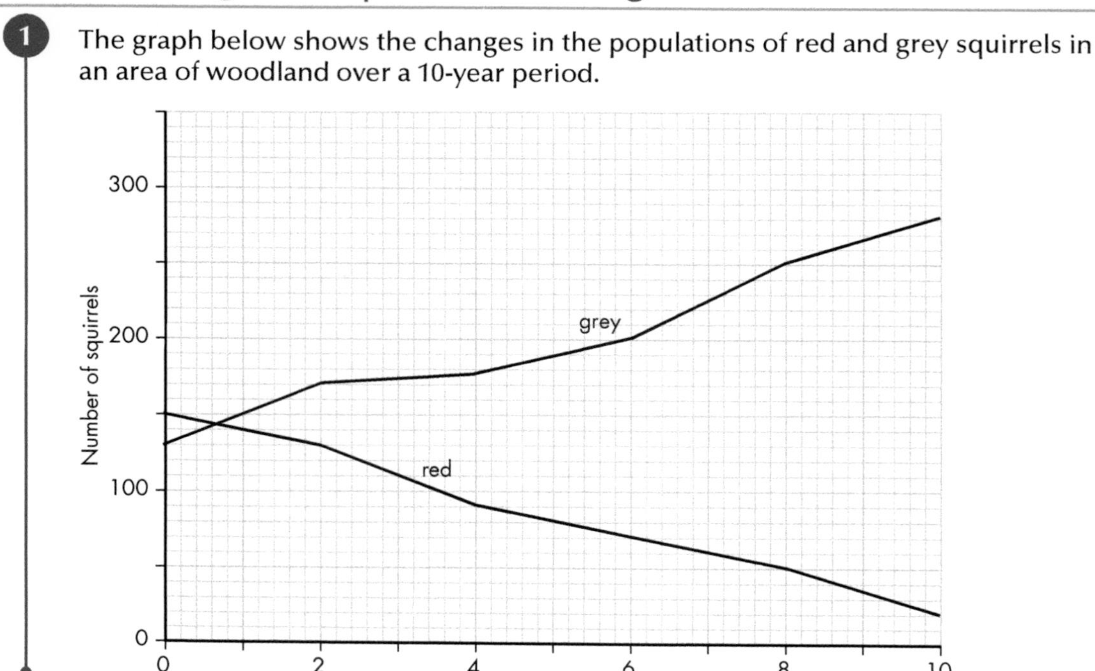

From the graph, the following conclusions were suggested.

1 The grey squirrel population increased by 150% over the 10-year period.

2 The red squirrel numbers decreased from 150 to 20 over the 10-year period.

3 After 8 years the grey squirrel population was five times greater than the red.

Which of the conclusions are correct?

A 1 and 2 only

B 1 and 3 only

C 2 and 3 only

D 1, 2 and 3

2 Conservationists did an observational study across different parts of Scotland, looking for the presence of the invasive species, American mink, and the non-native species of water vole. They observed five different geographical areas across Scotland, all next to large rivers, for one month.

The data collected is shown below.

Area	Number of organism spotted	
	American mink	Water vole
1	14	1
2	0	23
3	4	8
4	21	0
5	0	45

a Name the independent variable in this observation.

b Suggest how the conservationists could improve the design of the study to increase the validity of the results.

c Suggest how the conservationists could improve the study to increase the reliability of the results.

d Calculate the percentage increase in American mink from area 1 to 4.

Notes

Notes

Notes

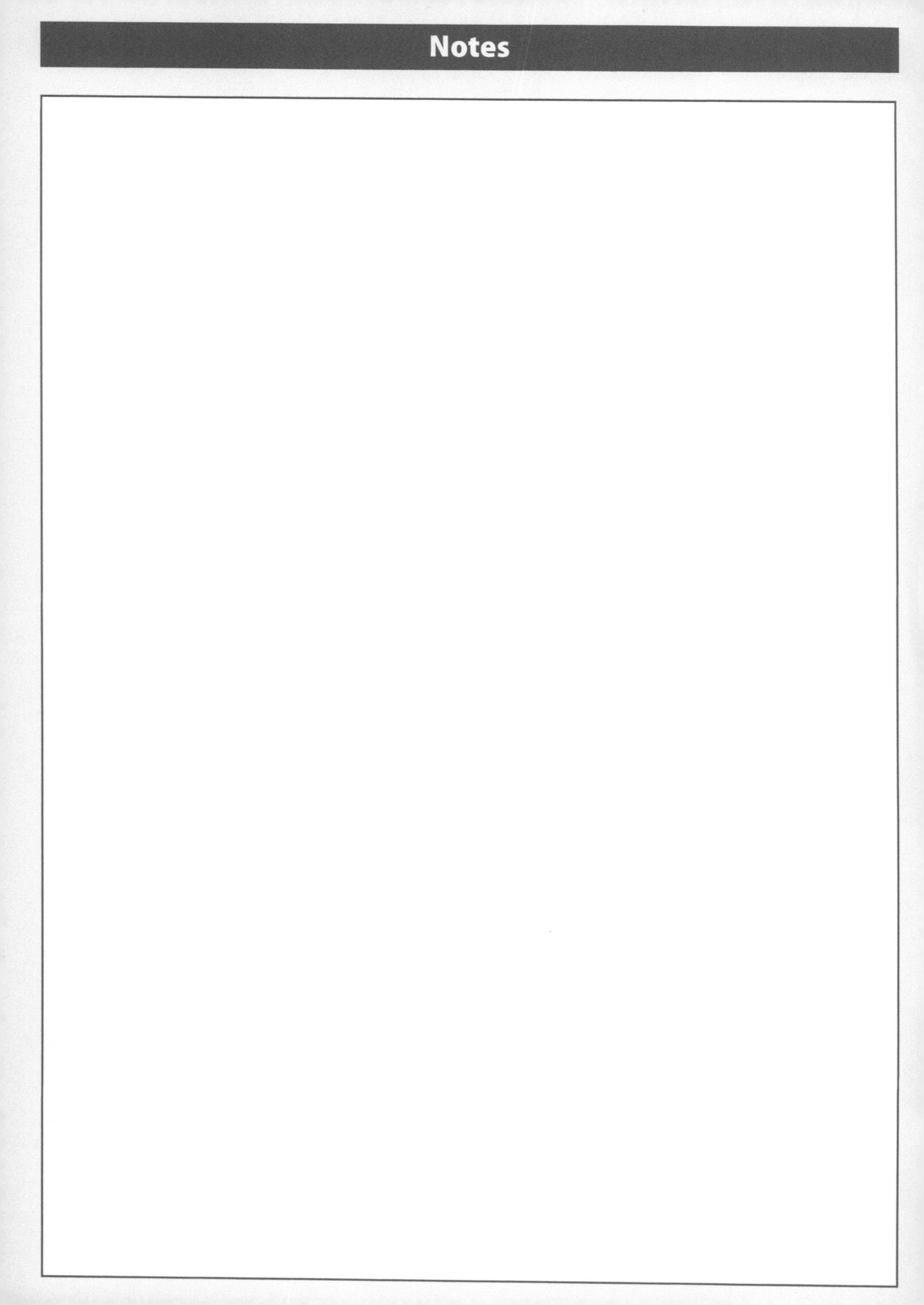
Notes

Notes

Notes